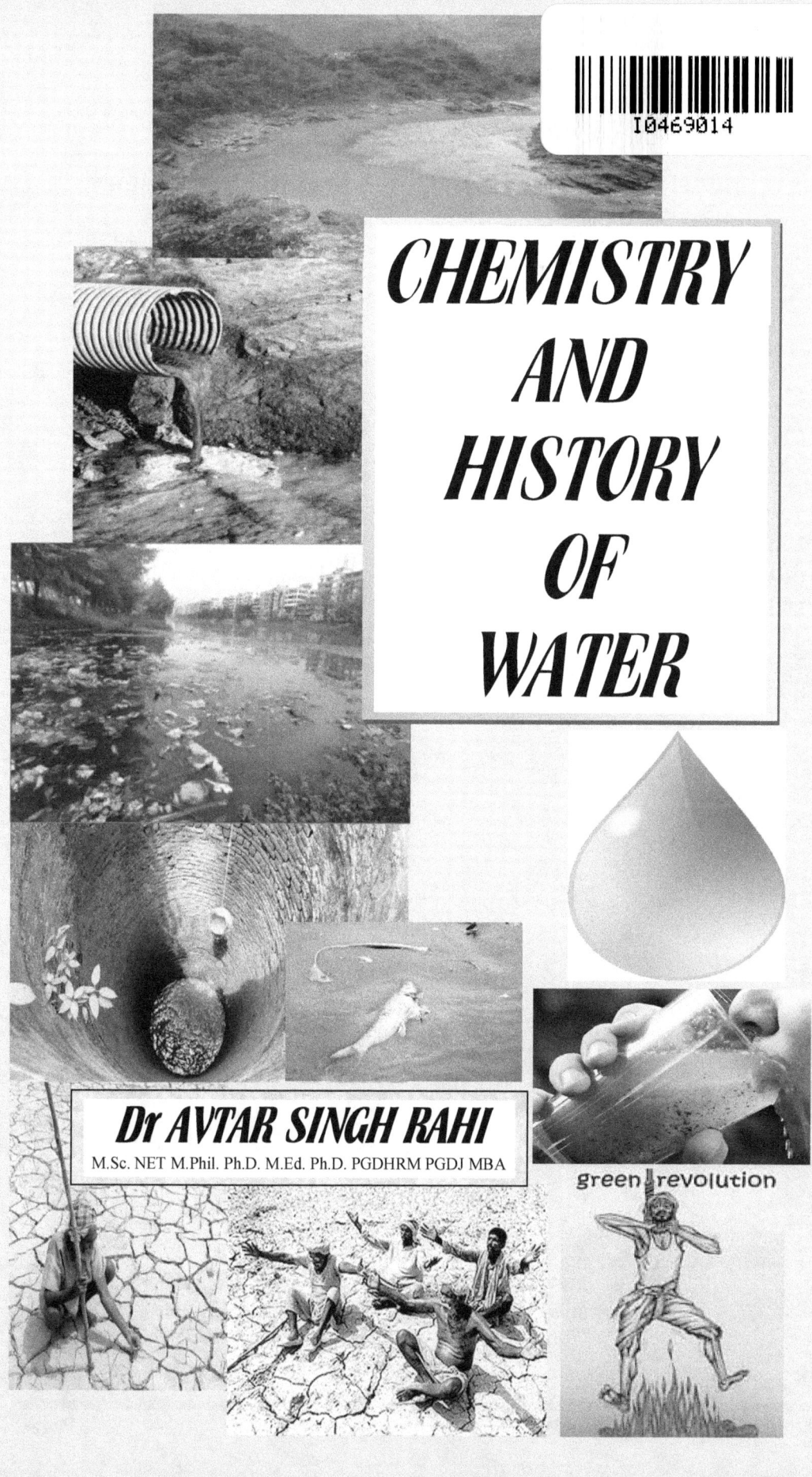

CHEMISTRY AND HISTORY OF WATER

Dr AVTAR SINGH RAHI

M.Sc. NET M.Phil. Ph.D. M.Ed. Ph.D. PGDHRM PGDJ MBA

green revolution

कैमिस्टरी एण्ड हिस्टरी
ऑफ वाटर
(पानी की स्थिति, गुण व
व्यवस्था)
डॉ अवतार सिंह राही

Chemistry and History of Water
(Availability, Properties and
Management of Water)
Dr Avtar Singh Rahi

ISBN-13: 978-1530698462
ISBN-10: 1530698464

संस्करण
Edition   **2011**

**THIS BOOK IS DEDICATED
TO THE 1.2 BILLION PEOPLE
WHO DO NOT HAVE ACCESS
TO SAFE AND CLEAN DRINKING WATER
WITH A PRAY
TO THE ALMIGHTY GOD
TO BLESS THE DON'T HAVES!**

CONTENTS

Chapter No.		Title	Pages

Chapter-1

Water is the most precious gift of the nature and one of the substances, essential for sustenance of life. It is one which influences economic, agricultural and industrial growth of the mankind. Several fundamental rights especially those guaranteeing the rights to food, health and development cannot be attained without guaranteeing access to clean water (Gleick, 1999). *Importance of water led the United Nations General Assembly to designate **22 March** of each year as the "WORLD DAY FOR WATER"* by adopting a resolution in 1992-United Nations Conference on Environment and Development (UNCED) in Rio de Janeiro.

Water is considered as one of the nutrients, although it yields no calories; it enters into structural composition of the cell and is an essential component of diet (Baloch et al., 2000). *It is not only the basic need for human existence but also a vital input for the development activities.* It is necessary for all metabolisms in the body and contributes to heat regulation by perspiration. Degradation of water quality creates water scarcity and limits its availability for human use and ecosystem and thereby impacts the optimum management of water resources (Rao and Mamatha, 2004). *Its availability in sufficient quantity and of right quality is a necessary infrastructure for promoting better quality of life.* A daily per capita consumption of two liters by a person weighing approximately 60Kg is generally assumed (WHO, 1996). Gleick (1999) estimates 50 lpcd as a true minimum to sustain life in moderate climatic conditions and average activity levels. Water has always been one of the most precious commodities. Water functions as a solvent for wide variety of chemical substances and facilitates industrial cooling and transportation. *Water has acted as the driving force of every civilization and is a part of all basic human needs, including food, drinking water, sanitation, health, energy and shelter. Without water we can have no society, no economy, no culture, and no life.* The Mohenjodaro and Harappan civilizations have thrown light on the fact that people of even that period had given importance to proper and efficient water supply management. Water is being used in all religious rituals and ceremonies since Ancient India because it is believed that the pure, divine and well provided waters convey the offerings to Gods.

Vedic writings speak of the value of water:
"Apo Nara Eti Prokta Apo Vai Nasvnv
Ta Yadsyayanm Purv Ten Narayana Smarth"
(Water is the child of the Almighty God)

Gurubani (Japji, Guru Nanak) makes a most reverential recitation:
"Pavan Guru Pani Pita Mata Dharti Mahat"
(Air is the Guru, Water the father and Earth the great mother)

Charaka Samhita Sutrasthanam (196) also states the importance of water:
"Jalamekam vidham sarvam patatyaindram nabhastalat,
Tatpatatpatitam caiva desakalavapeksate."

It is considered that Lord Indra directs the fall of water from heaven according to the activities performed by mortals. This water while falling from the sky acquires properties depending upon time and space (Krishnamurthy R, 1996).

The importance of water can be understood by the fact that many great civilizations in the past sprang up along or near water bodies in India and abroad. The development of water resources has often been used as a yardstick for socioeconomic and health status of many nations worldwide including India. Clean, fresh drinking water is essential to human and other life forms. *Chemically pure water does not exist for any appreciable length of time in nature.* Our natural environment supplies us clean drinking water. But while falling as rain, water picks up small amounts of gases, ions, dust, and particulate matter from the atmosphere. Then, as it flows over or through the surface layers of the earth, it dissolves and carries with it some of almost everything it touches, including that which is dumped into it by man. Provision of regular supply of clean drinking water is a birth right of all the citizens of a country. Access to safe drinking water has improved steadily and substantially over the last decades in almost every part of the world (Lomborg, 2001). Pure water means differently for different people. Homeowners are primarily concerned with domestic water problems related to color, odor, taste, and safety to family health. Chemists and engineers want to minimize scale deposition and pipe

Figure1.1:
Global distribution of the world's water
(Source: Rodda and Shiklomanov, 2003)

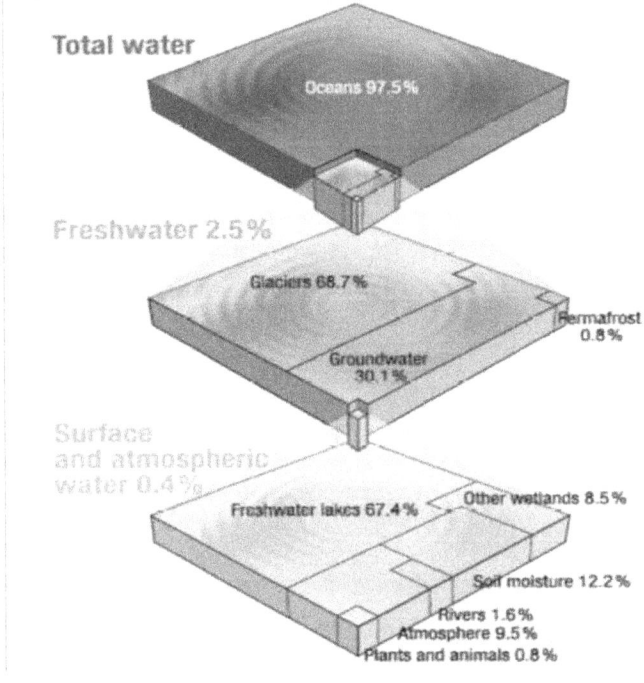

erosion. Farmers are interested in irrigationally (Chemical, Physical, and Osmotically) pure water. A correct balance in the sensory, physical, chemical and microbiological qualities of water makes it suitable for drinking. Water meeting these conditions is termed "Potable" meaning that it may be consumed in any desirable amount without concern for adverse effects on health (AWWA, 1990). Although roughly 66% of earth's surface is covered by water; water is scarce, as most is salt water. Less than 2.5% of all of the earth's water is fresh water; most of it is locked in polar ice caps and only a small fraction of that is available to supply the multitude of human uses.

Water is used in both productive and consumptive activities and contributes to rural and urban livelihoods in complex ways. *Crop and livestock production, agro-processing, fishing, ecosystems, recreation and human health are all influenced by the quality and quantity of available water.* Many people do not have access to enough water for both productive and consumptive uses. *About one-third of the world's population lives in countries that are experiencing moderate to high water stress resulting from increasing demands.* Water use has been growing at more than twice the rate of population increase during this century, and already a number of regions are water short. Today, 31 countries, accounting for less than 8% of the world's population, face chronic freshwater shortages. Among the countries likely to run short of water in the next 25 years are Ethiopia, India, Kenya, Nigeria and Peru. Parts of other large countries (e.g. China) already face chronic water problems (Tibbetts, 2000). According to Population Action International, based upon the UN Medium Population Projections of 1998, more than 2.8 billion people in 48 countries will face water stress or scarcity conditions by 2025. *By 2050, the number of countries facing water stress or scarcity could rise to 54,* with a combined population of four billion people - about 40% of the projected global population of 9.4 billion (Gardner-Outlaw and Engleman, 1997; UNFPA, 1997).

Water is difficult to treat, costly to transport and impossible to substitute and the consumption is increasing day by day due to increase in population. It is not only quantity that has to be preserved and ameliorated but also quality. Over one billion people each year are exposed to unsafe drinking water due to poor source water quality and lack of adequate water treatment. This results in 900 million cases of diarrhea each year (Rijal and Fujioka, 2001). The estimated number children that die each year due to water related diseases ranges from 2.5 million to 15 million (Burch and Thomas, 1998; Jorgensen *et al.*, 1998). *It is not considered uncommon, especially in developing countries, for defecation and urination to occur in rivers, lakes, and other bodies of water that are also used for domestic and recreational purposes* (Kloos *et al.*, 1997). World's people experience water scarcity as a result of rising and competing demands for water due to increasing population, the rapid expansion of irrigation, growth in industry and power generation, and due to lack of investments in infrastructure or capacity. The available water supply and its productive capacity are further strained by climatic change, land degradation, deterioration of quality, and the need to preserve environmental flows to protect aquatic and terrestrial ecosystems (Janmaat, 2004; Murgai et al., 2001; Postel, 1999).

Water supply sources may be surface water or ground water. Surface water is the term used to describe the water on land surface and it is produced by runoff of precipitation and ground water seepage. Water that seeps underground becomes "Groundwater" the major source of drinking water for many people. In fact, the bulk of the world's liquid fresh water is actually groundwater. Source water quality management is the first step in ensuring an adequate supply of safe drinking water, because surface water and ground water are treated differently under federal regulations, knowing the difference is important (AWWA, 1990). Groundwater quality management in many Asian countries is as serious as surface water. GEMS review of groundwater in Asia-Pacific (UNEP, 1996) mentioned the destruction, especially of shallow riverside and coastal aquifers, through over-pumping and pollution is greatly adding to the water crisis. The current status of water quantity and quality and related management issues in the Asia-Pacific regions are documented in a variety of publications, including an extensive series of monographs by ESCAP (e.g., 1994, 1995, and 1997), UN state of environment (1995), UNESCO publications, etc. The hydrological cycle works relatively quickly above ground, but slowly beneath it. It can take only a matter of months or years to recharge, and hence rehabilitate, surface waters, but groundwater recharge periods can be in the order of hundreds of years. *As a result, groundwater, once degraded, can be extremely difficult, sometimes impossible, to cleanse and restore.* Every glass of water we drink has, at least in part, already passed through fish, trees, bacteria, soil and many other organisms, including people. As it travels through these ecosystems, it is cleansed and made fit for human consumption. The undisturbed natural environment, with a few localized exceptions, provides water that is safe to drink in streams, lakes or wells. This supply of water is a "service" (benefit to humans) that the environment provides. Biodiversity is what underpins the ability of nature to provide this service by sustaining the continuous recycling of water, through the hydrological cycle. *We cannot properly preserve our water resources without first understanding how water circulates throughout the environment.*

1.1   HYDROCHEMISTRY AND POLLUTION

Hydrochemistry has a potential use for tracing the origin and history of water. The hydrochemistry can also be of immense help in yielding information about the environment through which water has circulated. Hydrochemistry can be helpful in knowing about residence times, flow paths and aquifer characteristics as the chemical reactions are time and space dependent. It is essential to study the entire system like atmospheric water (rainwater), surface water and groundwater simultaneously in evaluating their hydrochemistry and pollution effect. Pollutants in the atmosphere can be transported long distances by the wind. These pollutants are mostly washed down by precipitation. Composition of rainwater is determined by the source of water vapors and by the ion, which are taken up during transport through the atmosphere. In general, chemical composition of rainwater shows that rainwater is only slightly mineralized with specific electrical conductance (EC) generally below 50 μS/cm, chloride below 5 mg/L and HCO_3^- below 10 mg/L. Among the cations, concentration of Ca, Mg, Na & K vary considerably but the total cations content is generally below 15 mg/L except in samples contaminated with dust. The concentration of sulphates and nitrates in rainwater may be high in areas near industrial hubs. *Surface water is found extremely variable in its chemical composition due to variations in relative contributions of ground water and surface water sources. The chemical composition of groundwater vary depending upon several factors like frequency of rain, which will leach out the salts, time of stay of rain water in the root-zone and intermediate zone, presence of organic matter etc.* The movement of percolating water through larger pores in earth is much more rapid than through the finer pores. Apart from this, there are several other reactions including microbiological mediated reactions, which tend to alter the chemical composition of the percolating water. The overall effect of all these factors is that the composition of ground water varies from time to time and from place to place. Contaminants can be in the form of microorganism that barely visible in unaided eyes. *A number of authors have reported a statistically significant deterioration in the microbiological quality of water between the source and point of use in the home* (Welch et al., 2000; Genthe and Strauss, 1997; Simango et al., 1992). Coliforms are routinely found in diversified natural environments, as some of them are of telluric origin, but drinking water is not a natural environment for them. As a result, their presence in drinking water must be considered as harm to human health.

1.1.1 DISSOLVED CONSTITUENTS

The concentration of dissolved constituents in groundwater is comparatively higher than surface water. *It is common thinking that the potable water has few numbers of dissolved inorganic constituents but the real picture is quite different*, as about 58 elements have been enlisted in the literature. For example, the relative abundance of dissolved solids in potable water is classified into three categories:

i)   Major Concentration (1.0 to 1000mg/L): Sodium, Calcium, Magnesium, Bicarbonate, Sulphate, Chloride and Silica.

ii)   Secondary Concentration (0.01 to 1.0 mg/L): Iron, Strontium, Potassium, Carbonate, Nitrate, Fluoride and Boron.

iii)   Minimum Concentration (0.0001 to 0.01 mg/L): Antimony, Aluminum, Arsenic, Barium, Cadmium, Chromium, Cobalt, Copper, Germanium, Iodide and Zinc.

Everyone knows that water sustains life but not everyone knows that water endanger life as ingestion or exposure to contaminated water can cause several health hazards. Therefore, safe drinking water is essential to good health. In addition, diseases caused through consumption of contaminated water, and poor hygiene practices are the leading cause of death among children worldwide, after respiratory diseases (WHO, 2003).

1.1.2   WATER IMPURITIES

Water impurities include dissolved and suspended solids. Calcium bicarbonate is a soluble salt. A solution of sodium bicarbonate is clear because the calcium and bicarbonate are present as atomic sized ions, which are not large enough to reflect light. Some soluble minerals impart colour to the solution. Soluble iron salts produce yellow or green solutions; some copper salts form intensely blue solutions. Dissolved and suspended solids are present in most surface water. Suspended solids are substances that are not completely soluble in water are present as particles. These particles usually impart visible turbidity to water. The common impurities present in natural water may be classified as follows:

1. **Dissolved impurities:** The inorganic salts which dissolve in natural water are usually bicarbonates, chlorides, sulphates and nitrates of sodium, potassium, magnesium, calcium, aluminum and iron. Zinc and copper salts are also sometimes present in traces.
2. **Gases:** Several gases such as carbon dioxide, oxides of nitrogen and sulphur, hydrogen sulphide and ammonia may dissolve in natural water. These gases are present as pollutants in the atmosphere and dissolve in water during the rain fall.
3. **Organic matter:** huge amount of domestic sewage and industrial wastes are thrown into canals and rivers every year. These wastes contain organic compounds which contaminates the natural resources of water.
4. **Suspended impurities:** The surface water contains suspended particles of sand, slit and minerals eroded from the land.
5. **Pathogenic microorganisms:** Various pathogenic microorganisms such as bacteria, viruses etc. also enter in to water bodies through sewage and other wastes.

1.2   SOURCES OF POLLUTION

Many factors contribute to the quality of water. At a basic level, water hydrology and functions are dependent on five variables: Climate, Geology, Soils, Land Use and Vegetation (Morisawa and LaFlure, 1979). Of these variables, land use and vegetation are the only variables over which man has direct control, underscoring their primary significance in the land use planning process. Booth and Jackson (1997) identify changes in upland land use as critical in determining, overall water function, degradation, and rehabilitation potential and found that even with best efforts, some downstream aquatic system damage is probably inevitable. *Many developing regions suffer from either chronic shortages of freshwater or the pollution of readily accessible water resources* (Lehloesa & Muyima, 2000). Trace metals have been referred to as common pollutants, which are widely distributed in the environment with sources mainly from the weathering of minerals and soils. However, the level of these metals in the environment has increased tremendously in the past decades as a result of inputs from human activities (Prater, 1975). Pollution of surface and groundwater resources occurs through Point and Diffuse sources. Examples of point source pollution are effluents from industries, sewage-treatment plants and untreated domestic sewage. The main sources of diffuse pollution may be anthropogenic activities, such as agricultural applications of fertilizers and pesticides or of geo-chemical origin, such as natural contamination of groundwater sources by fluoride, arsenic and dissolved salts. Discharges from point sources as well as from land application facilities tend to have higher dissolved solids than natural surface waters (Kent and Belitz, 2004). Pollution from point sources can be controlled by disposal in engineered facilities, treatment and recycling of waste materials. Minimizing application of fertilizers and pesticides is a way to control pollution from agricultural activities. Natural contamination of groundwater sources by fluoride, arsenic and dissolved salts is dealt with by suitable treatment of extracted groundwater.

1.2.1   Pollution from point sources

1.2.1.1 <u>Industrial pollution:</u>

In case of industrial units, effluents in most of the cases are discharged into pits, open ground, or open unlined drains near the factories, thus allowing it to move to low lying depressions resulting in groundwater pollution. *The industries, which are burgeoning at a fast rate, produce several million m^3 of wastewater per day, which is discharged into rivers and streams.* Thus the magnitude of damage caused to our water resources can be estimated from the fact that about 70% of rivers and streams in India contain polluted water. The incidence of surface and groundwater pollution is highest in urban areas where large volumes of waste are concentrated and discharged into relatively small areas. The groundwater contamination is detected only some time after the subsurface contamination begins. *Although the industrial sector accounts for only 3% of the annual water withdrawals in India, its contribution to water pollution, particularly in urban areas, is considerable.*

1.2.1.2 Pollution from domestic activities:

Inadequate treatment of human and animal wastes contributes to the high incidence of water-related diseases in the country. To date, only 14% of rural and 70% of urban Indian inhabitants have access to adequate sanitation facilities. Therefore, *water contaminated by human waste is often groundwater table from faulty septic tanks or pit latrines.* The level of faecal coliform bacteria in most rivers often exceeds the WHO (World Health Organization) standards and is responsible for causing a number of gastrointestinal ailments among the population. All of India's 14 major river systems are heavily polluted, mostly from the 50 million cubic meters of untreated sewage discharged into them each year. The domestic sector is responsible for the majority of the wastewater generation in India. Combined, the 22 largest cities in the country produce over 7267 million litres of domestic wastewater per day.

1.2.2 Diffuse pollution

The main sources of diffuse pollution may be anthropogenic activities, such as agricultural applications of fertilizers and pesticides or of geo-chemical origin, such as natural contamination of groundwater sources by fluoride, arsenic and dissolved salts.

1.2.2.1 Agricultural activities:

In agriculture (Ongley, 1998), there is rarely interaction between agricultural managers and water quality managers with the result that the database that is needed to assess agricultural impacts on water quality is never available. *Use of fertilizers and pesticides to improve soil fertility and crop protection has created an environmental menace.* Both these products find their way into the food chain and have implications on human health. Fertilizers and pesticides have entered the water supply through runoff and leaching to the groundwater table and pose a hazard to human, animal and plant populations. Some of these chemicals such as hexachlorocyclohexane (HCH), dichlorodiphenyltrichloroethane (DDT), endosulfan, methyl malathion, malathion dimethoate, etc. are considered as extremely hazardous by the WHO and are banned or are under strict control in developed countries.

1.2.2.2 Geological origin:

Pollution of groundwater resources due to geological conditions has become a matter of serious concern (Rao and Mamatha, 2004). Groundwater in certain geological formations may not be of desired quality for potable use because of geo-chemical conditions. Most of the critical quality-related problems of ground water in India are cited as geogenic largely due to major inorganic pollutants like fluoride and arsenic. Arsenic problem prevails in 3136 habitations and fluoride is endemic in 36,988 habitations (DDWS, 2004). Arsenic contamination of groundwater invariably arises from natural geological and environmental conditions. Arsenic arises in many ores and minerals and is frequently present in combination with iron and manganese oxides; under various natural conditions it can be rendered soluble and released into the groundwater. Groundwater with high fluoride content is found mostly in calcium-deficient groundwater in many basement aquifers, such as granite and gneiss, in geothermal waters and in some sedimentary basins. Twenty states (about 200 districts) in India have been identified as endemic to fluorosis due to abundance in natural occurring fluoride-bearing minerals. The fluorosis problem is severe in India as almost 80% of the rural population depends on untreated groundwater for potable water supplies (Rao and Mamatha, 2004). Factor contributing to excess fluoride in groundwater in rural regions of India is the over-exploitation of groundwater resources for agricultural and drinking water purposes. The quantum of water drawn from the aquifers exceeds aquifer re-charge that aids the concentration of fluoride in the aquifers. Though iron content in drinking water may not affect the human system as a simple dietary overload, in the long run prolonged accumulation of iron in the body may result in homo-chromatosis, where tissues are damaged. A total of 106,019 sq km area (about 31%) of Rajasthan comes under saline groundwater. Arsenic in groundwater has been reported in shallow aquifers from 61 blocks in eight districts of West Bengal.

The impact of anthropogenic activities has been so extensive that the water bodies have lost their self-purification capacity to a large extent (Sood et al., 2008). Over one billion people in the world lack access to safe water supplies. Exposure to environmental health risks in early childhood leads to permanent growth faltering, lowered immunity and increased mortality. Poor water, sanitation, hygiene and inadequate water resources management, account for half of the causative factors behind childhood and maternal underweight and hence child growth (World Bank 2008). Approximately 3.1% of deaths (1.7 million) and 3.7% of disability-adjusted-life-years (54.2 million) worldwide are attributable to unsafe water, poor sanitation and hygiene (WHO, 2005). *Currently, it is estimated that two thirds of people in Asia still lives without any access to drinking water* (WHO, 2004). According to WHO (1997), in most of the countries, the principal risk to human health associated with the consumption of polluted water are microbial in nature (although the importance of chemical contamination could not be underestimated). Although the existence and dangers of pathogenic microbes in surface drinking water supplies have been recognized since long, groundwater supplies, hence well and springs were generally thought to be naturally protected. This assumes that protection was attributed to the natural filtration and neutralizing properties of subsurface soil and geological strata (Robertson and Edberg, 1997). Contamination in surface and ground water can range from natural substances leaching from soil. Water quality deterioration may occur due to sources of faecal pollution including grazing cattle, natural animal population, septic tanks, failed sewage systems, recreational users, and summer storm activities etc. (Crabill et al., 1999).

Faecal pollution of drinking water causes water-borne diseases, which wiped out entire population of cities (Farah *et al.*, 2002). The coliform groups of bacteria principally infect water used for domestic, industrial or other purposes (Zamaxaka *et al.*, 2004). Consequently, this has caused many people to suffer from various diseases (Tanwir *et al.*, 2003). Unsanitary disposal of refuge and garbage, increased use of agricultural pesticides and fertilizers, industrial operations, use of pit latrines and problems with septic tank systems constitute major anthropogenic activities causing groundwater pollution (Baloch et al., 2000; Sichingabula and Nkhuwa, 1998; Knox and Canter, 1996; Koppe, 1973). Freedom from contamination with faecal matter is the most important parameter of water quality because human faecal matter is generally considered to be a greater risk to human health as it is more likely to contain human enteric pathogens (Scott et al., 2003). Groundwater supplies have some advantages over surface water. Data have shown that groundwater is less susceptible to favour bacterial growth (Niquette et al., 2001). Groundwater supply can be easily developed at a small capital cost (Raghunath, 1982), so preferred especially in rural areas.

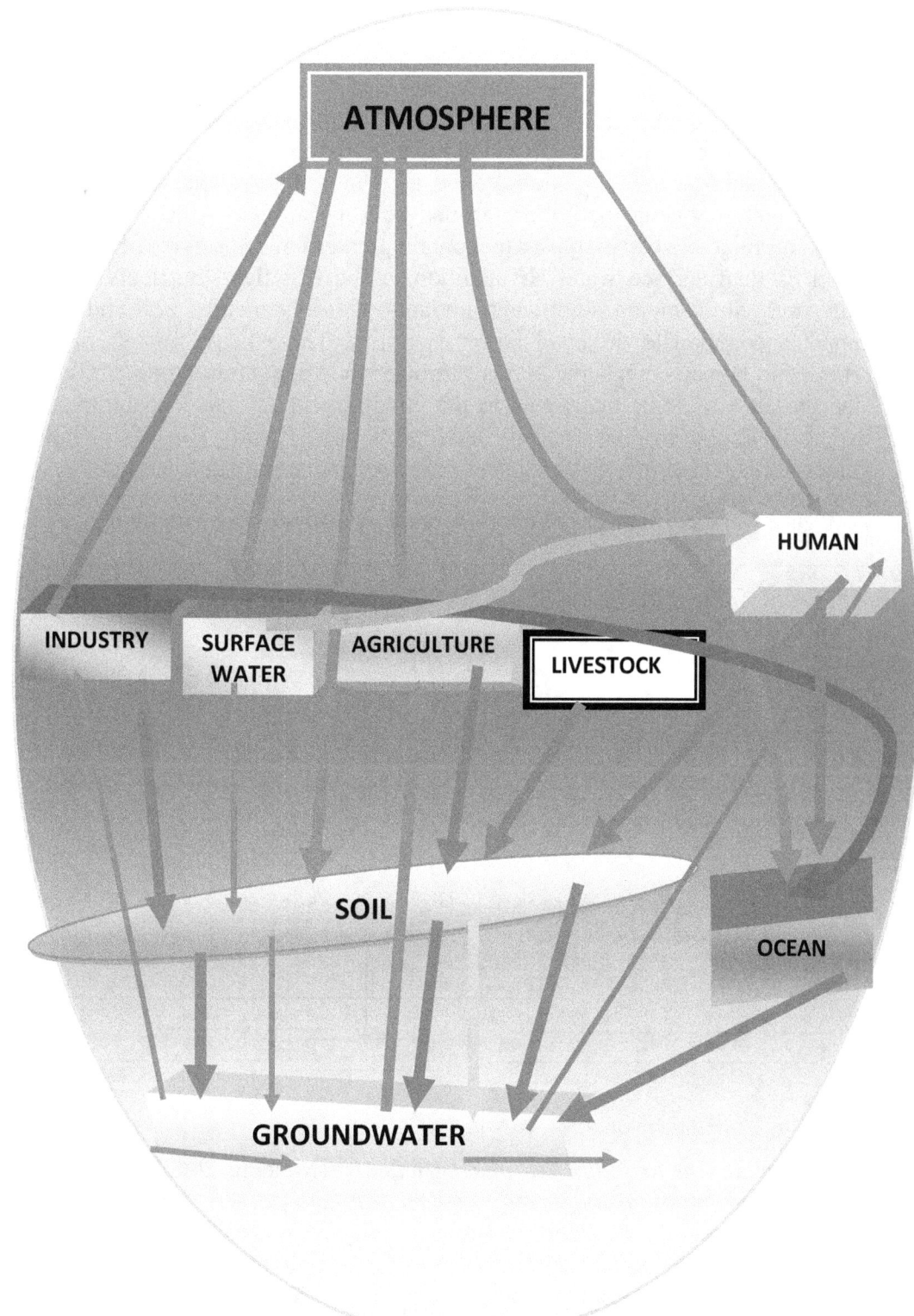

Figure 1.2: Groundwater Pollution – Diagrammatical Representation

1.3   SOIL - THE ULTIMATE SINK

Soil is considered an ultimate disposal sink, but soil profile is also responsible for water pollution. Additional constituents might be added to ground water by leachate from solid waste dumping sites (Kumar, 2008; Robinson and Maris, 1985; Nicholson et al., 1983; Olaniya and Saxena, 1977). Besides anthropogenic activities, groundwater because of long contact with rocks and mineralized soils, usually contain greater concentrations of inorganic additives than surface water. In addition to the beneficial nutrients, sewage water may also contain significant amounts of heavy metals. Soil and plants cannot accommodate these all harmful entities. These heavy metals may be returned to human beings and animals through food chain (Gulfraz et al., 1997). The presence of toxic chemicals in the environment at low concentrations is undetectable and produces harmful effects. Sewage usually consists of higher values of pH, alkalinity, TDS, nitrites, nitrates, conductivity and cations.

Figure 1.3: Pictorial Presentation - Sources of Groundwater Pollution
Source: Foster et al., 2002

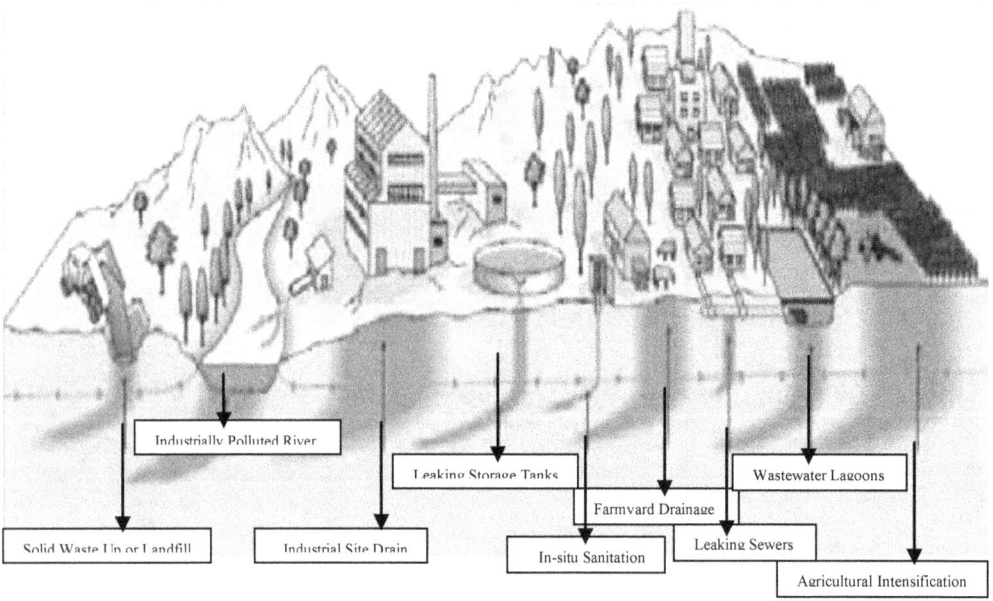

Investment in Industries and agriculture (Molden, 2007) *has made a positive contribution to livelihood, food security and poverty reduction, but its negative effects are unforgettable.* In the vicinity of industrial houses in almost all cities of India, surface and ground water is polluted due to dumping of industrial wastages. Approximately 70% of fresh water is consumed by agriculture (Baroni et al., 2007). Humans change almost all aspects of the hydrological cycle and the ecosystem of which it is part; shifting water around for different uses, overusing it and degrading the environment that supplies it. In fact, the rate of loss of biodiversity from freshwater ecosystems is the fastest of all biomes. Poor access to drinking water is in most cases a direct result of human behaviour. Unregulated groundwater extraction for agriculture and industrial use is emerging as a major threat to resource sustainability, threatening drinking water supplies. Contamination of groundwater through unregulated mining, the use of chemical fertilizers and salinity ingress in coastal areas is becoming alarming. *Many of irrigation's negative environmental impacts arise from the diversion of water away from natural aquatic ecosystems, such as rivers, lakes, oases and other groundwater dependent wetlands.* The direct and indirect negative impacts have been well documented, including salinization, channel erosion, decline in biodiversity, introduction of invasive alien species, reduction of water quality, genetic isolation through habitat fragmentation and reduced production of floodplains and inland and coastal fisheries (Richter et al., 1997; Bunn and Arthington, 2002; Pimentel et al., 2004; Khan et al, 2006; Gordon et al., 2010). Investments in biofuels and other projects under the clean development mechanism may reduce the rate of climate change, but will also have significant impacts on water and land (Zomer et al., 2006).

The most important pollutants are probably chloride and nitrate, which can be derived from leaking sewers, landfills, un-sewered sanitation, livestock farming and agricultural fertilizers. High chloride concentrations may also indicate saline intrusion in coastal areas or the use of salt for road de-icing in cold climates. These pollutants can be indicators of impact from both rural and urban activities. Nitrate in particular can be problematic, as un-sewered sanitation and agriculture often occur in close proximity. Since these parameters cannot be removed or eliminated by means of water disinfection and can bring about a lot of health hazards, a complementary unit must be added to water treatment processes or these resources should be considered as undrinkable water resources (Chapman, 1996).

1.4   POLLUTION THREATS TO GROUNDWATER IN RURAL AREAS

Irrigation agriculture has been credited since long with raising farm income and reducing rural poverty (Hussain and Hanjra, 2004), though irrigation schemes have also been responsible of large scale epidemics of vector borne diseases like malaria, and schistosomiasis (Oomen et al., 1990). *History have shown that irrigation based societies are fragile and water scarcity, soil salinization, and conflict over (river) water have led to the decline of once powerful and advance societies like Mohenjo-Daro, Mesopotamia and Egypt* (Postel, 1999). Intensification of agriculture, increasing productivity, rapid extension of irrigation, fertilizer application and pest control, put unanticipated adverse impacts on the quality of underlying groundwater. Microbiological health risks remain associated with many aspects of water use, including drinking water in developing countries, irrigation reuse of treated wastewater and recreational water contact (Grabow, 1991). The information on the quality of groundwater yields valuable knowledge regarding their possible effects on physicochemical properties of the soil and its productivity (Sharma and Minhas, 2004). Several factors may be considered while assessing the agricultural pollution to groundwater:

- ❖ Area covered in cultivation is very large, often equal to the aquifer, so can potentially lead to widespread pollution of the groundwater, because the total loadings may be high enough to bring considerable changes and concentrations may significantly exceed drinking water guidelines values.
- ❖ In many developing countries water supply and sanitation are unable to keep pace with urbanization and municipalities are therefore forced to set priorities with the use of waste water in agriculture (Scott et al., 2004) which once considered useful because of nutrient concentration (van der Hoek et al., 2002) but is a father of many diseases.
- ❖ Use of waste water in agriculture increases nitrate contamination of groundwater (Chilton et al., 1998), salts, metals and other contaminants in soil and threaten agricultural sustainability (Chang et al., 2002).
- ❖ Improper crop management may change the status of nutrients in soil, leading to deficiency.
- ❖ External compensation of deficient minerals can alter the aquifer qualities.
- ❖ Irrational extraction of water, without recharge or insufficient recharge increases salinisation, making it unfit.
- ❖ Blind use of fertilizers, pesticides and over-productivity can easily create waste and barren lands.
- ❖ Leaching from unprotected dug-wells and ponds in rural areas is also a rich source of contamination.
- ❖ Pollution from human settlements lacking appropriate sanitary infrastructure.
- ❖ Downward leakage of the wastes of Livestock and humans, and irrigation by wastewater may exceed the contamination in shallow aquifers significantly.
- ❖ Nearby industrial units or drainage carrying industrial wastes may further deteriorate the situation.
- ❖ If the problem persists, downward leakage of shallow aquifers may cause contamination of the deeper aquifers in the long term.
- ❖ Slow and delay in treatment and lack of proper water management in rural areas may further aggravate the water contamination by reaction products, byproducts, etc.

1.5   STATUS OF INDIA

By its nature and multiple uses, water is termed as a complex subject. Although water is a universal and global issue, the problems and solutions are often highly localized. With 2.4% of land and 4% of water resources, India has to support 16% of world's population and 15% of livestock. *Behavior of groundwater in the India subcontinent is highly complicated due to the occurrence of diversified geological formations with considerable litho-logical and chronological variations, complex tectonic framework, climate-logical dissimilarities and various hydro-chemical conditions.* Broadly two groups of rock formations have been identified, depending on characteristically different hydraulics of ground water viz. Porous Formations and Fissured Formations. Porous formations include areas covered by alluvial sediments of river basins, coastal and deltaic tracts and narrow valleys or structurally faulted basins, while Fissured formations occupy almost two-thirds of the country and include igneous and metamorphic rocks excluding volcanic and carbonate rocks (e.g. Granites, gneisses, charnockites, khondalites, quartzites, schists and associated phyllites, slates, etc.), volcanic rocks (e.g. Basaltic lava flows of Deccan Plateau), consolidated sedimentary rocks excluding carbonate rocks (e.g. Rocks occur in Cuddapahs, Vindhyans and their equivalents) and Carbonate rocks (e.g. Limestone in the Cuddapah, Vindhyan and Bijawar groups of rocks).

While India is considered rich in terms of annual rainfall and total water resources, its uneven geographical distribution causes severe regional and temporal shortages. India's rivers carry 90 percent of the water during the period from June-November. Thus, only 10 per cent of the river flow is available during the other six months.

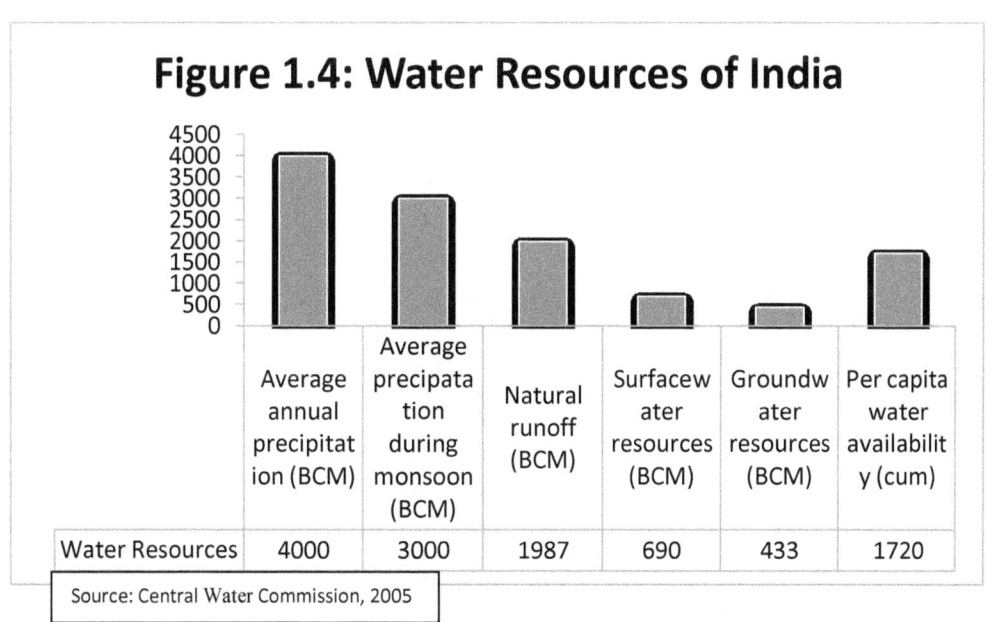

Figure 1.4: Water Resources of India

	Average annual precipitation (BCM)	Average precipatation during monsoon (BCM)	Natural runoff (BCM)	Surfacewater resources (BCM)	Groundwater resources (BCM)	Per capita water availability (cum)
Water Resources	4000	3000	1987	690	433	1720

Source: Central Water Commission, 2005

India is struggling hard to provide its citizens with all basic amenities, but clean drinking water is not available to great number of people mainly because of rising population, rising levels of pollution in the environment, poor upkeep of water supply lines and faulty drainage system. Groundwater is the major source of water supply for drinking and other purposes in the rural areas of India (Gupta et al., 2009). In India ponds, rivers and groundwater are used for domestic and agricultural purposes (Bhandari and Nayal, 2008). India is now the biggest user of groundwater for agriculture in the world. Groundwater irrigation has been expanding at a very rapid pace in India since the 1970s (Vijay Shankar et al., 2011). Data from Minor Irrigation Census (2001) showed that *three states (Punjab, Uttar Pradesh and Haryana) accounted for 57% of the tube-wells in India.* On an average, there were 27 tube-wells per square kilometer of net sown area in Punjab. Data from the National Sample Survey Organization (1999) indicates that dependence on groundwater as the principal source (tube well/hand pump) has increased both in rural and urban areas. About three-fourth of rural population is relying on groundwater sources as shown in figure.

Figure 1.5a: Percentage Distribution of Households by Principal Source of Drinking Water in India: 1988-1998 (Urban and Rural areas)
(Source: National Sample Survey Organization, 1999)

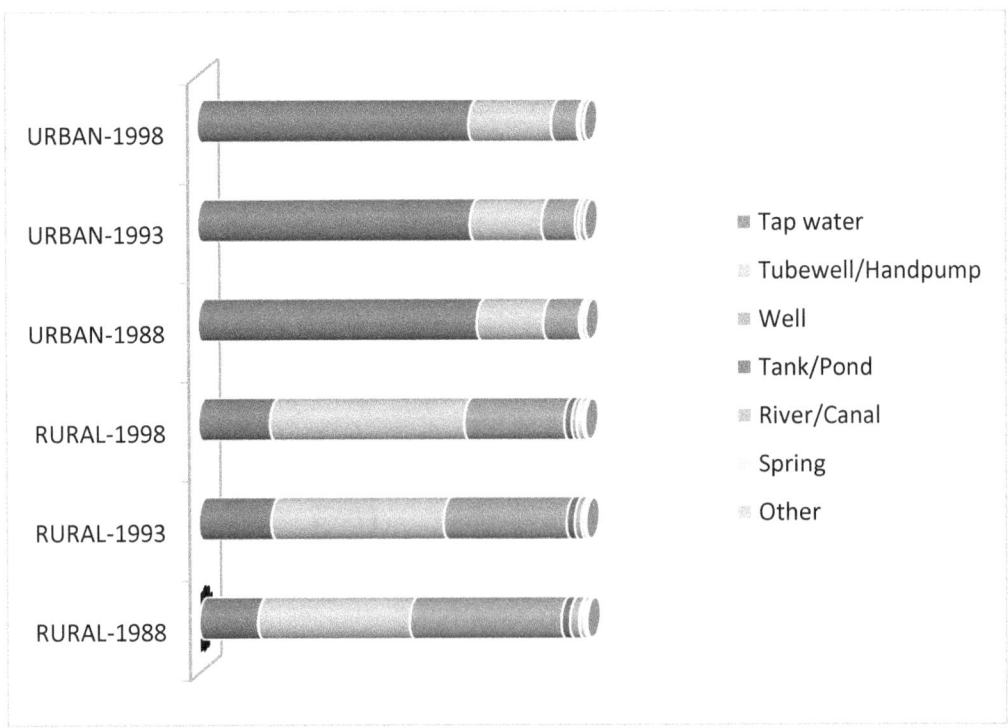

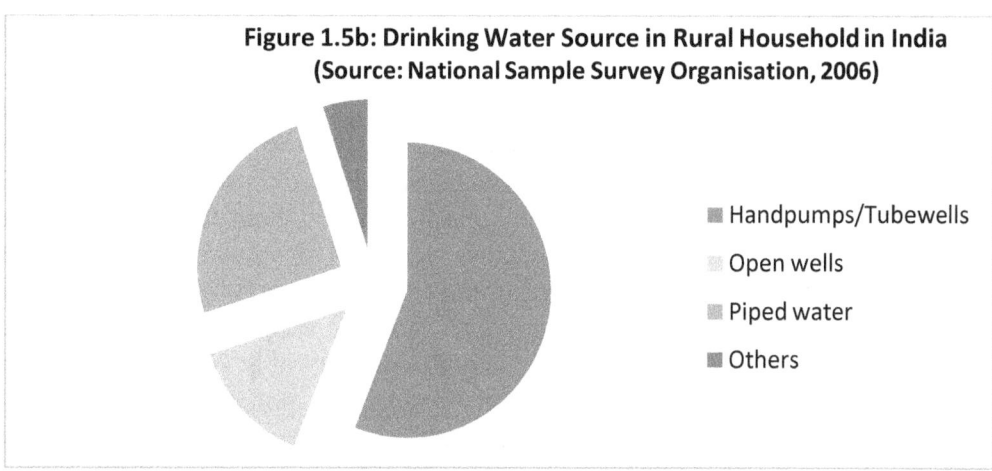

Figure 1.5b: Drinking Water Source in Rural Household in India
(Source: National Sample Survey Organisation, 2006)

- Handpumps/Tubewells
- Open wells
- Piped water
- Others

By far the most important driver in water use during the coming decades will be the increase and changes in global food demand due to population growth and changes in diet (Fraiture and Wichelns, 2010). The available utilizable water resource of the country is considered insufficient to meet all future needs (CWC, GOI, 2005). *India's finite and fragile water resources are stressed and depleting, while sectoral demands (including drinking water, industry, agriculture, and others) are growing rapidly.* Rapid population growth, urbanization and industrialization have led to a greater demand for an increasingly smaller supply of water resources in the country. Of the present water usage in the country, majority is consumed in agriculture (70-90%), and the remaining is consumed in industrial activities and for domestic purposes like drinking water and sanitation (Rao et al., 2004). To meet the food requirement of the estimated population of India in 2025, agriculture will need increased and extensive application of agro-chemicals and 731 to 867 Bm^3 of irrigation water, while the utilizable water resources of the country may remain in between 219 to 355 Bm^3 in 2025.

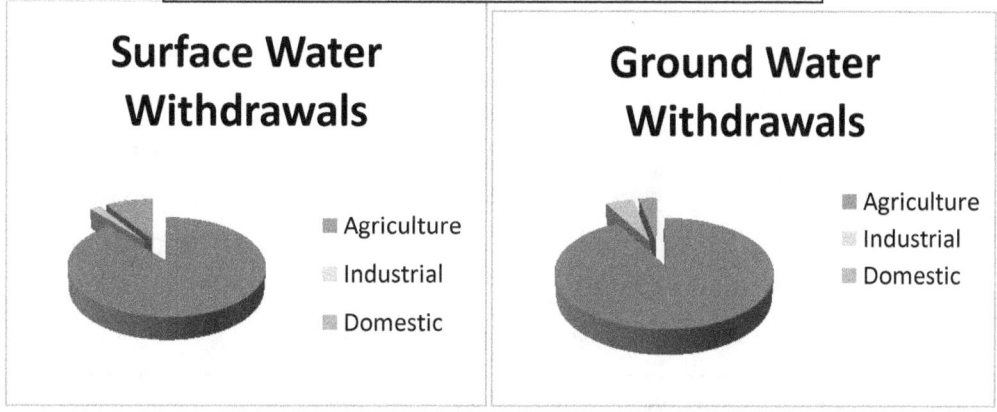

Figure 1.6: Water Usage by Sector
Source: Earth Trends 2001, World Resources Institute

Surface Water Withdrawals

- Agriculture
- Industrial
- Domestic

Ground Water Withdrawals

- Agriculture
- Industrial
- Domestic

Problems with water quality are often as severe as problems with water availability. Slow growth of surface canal irrigation and watershed projects induced rapid growth of groundwater irrigation especially after 1970 (Nayak, 2009). While the technology has allowed drinking water to be pumped from the ground through bore-wells and hand-pumps, it also provided irrigation sector the means for unfettered pumping of groundwater through millions of irrigation bore-wells (nearly 3.7 million in 2004), leading to imbalance in natural ecological system resulting scarcity and pollution of ground water (Daw, 2004). Over-extraction of groundwater is leading to alarming depletion of groundwater aquifers in several areas (Iyer, 2003; Reddy, 2004). It is observed that the irrigation potential created has exceeded the ultimate potential, showing that mining of groundwater is taking place through exploitation beyond the dynamic resources in many states (Government of India, 2007). Cheap and un-metered electricity, slow development of surface irrigation, and poor management of canal systems encouraged groundwater development. Over the last two decades, 84 percent of the total addition to net irrigated area came from groundwater, and only 16 percent from canals (Brisco and Malik 2006).

Figure 1.7a: Sources of Irrigation in India: 1950-51 to 2001-02 ('000' Hectares)
(Source: Briscoe and Malik, 2006)

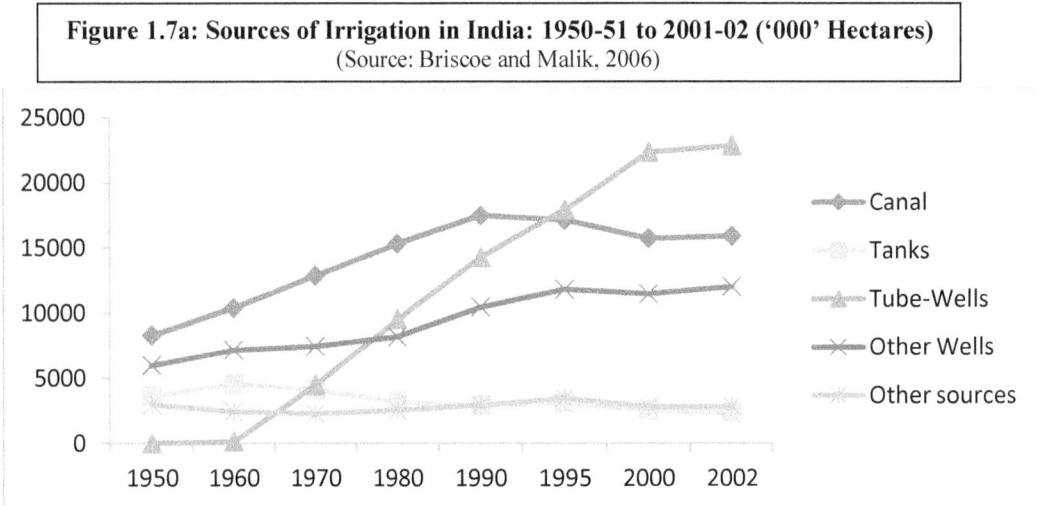

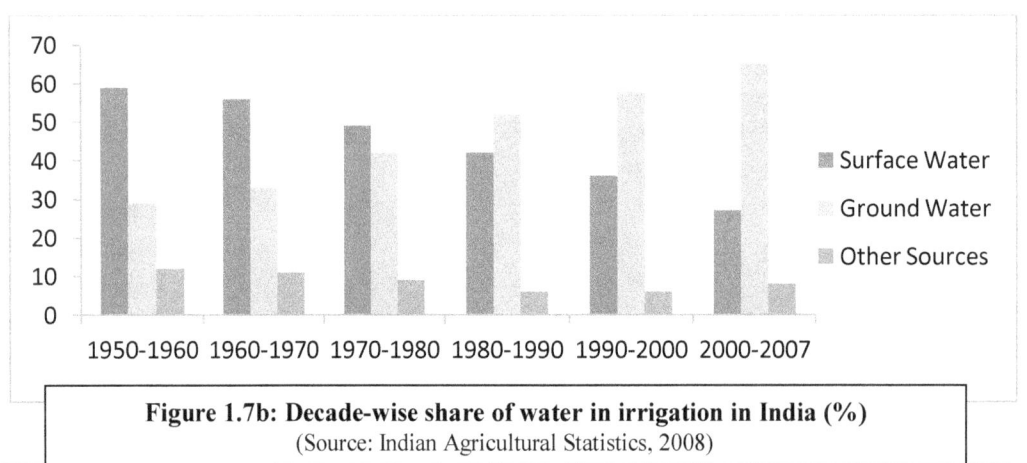

Figure 1.7b: Decade-wise share of water in irrigation in India (%)
(Source: Indian Agricultural Statistics, 2008)

This expansion of groundwater use resulted in speedy decline in the groundwater table in several parts of the country. Out of 4272 blocks in the country (except Andhra Pradesh, Gujarat and Maharashtra where ground water resource assessment has been carried out on the basis of mandals, talukas and watersheds respectively), 231 blocks have been categorized as "Overexploited" where the stage of ground water development exceeds the annual replenishable limit and 107 blocks are "Dark" where the stage of ground water development is more than 85%. Besides, 6 mandals have been categorized as "Overexploited" and 24 as 'Dark' out of 1104 mandals in Andhra Pradesh. Similarly out of 184 talukas in Gujarat, 12 are "Overexploited" and 14 are 'Dark' and out of 1503 watersheds in Maharashtra, 34 are 'Dark'. Problem villages have gone up to more than 200,000 in 2004 and habitations with problems of inadequate water availability are stated at another 300,000.

Water pollution is a serious problem in India as almost 70% of its surface water resources and a growing number of its ground water reserves are already contaminated by inorganic, organic and biological pollutants. High extraction of ground water due to increased demand has given rise to compounded Arsenic and Fluoride contamination and Salinization.

Figure 1.8: State-wise Groundwater Quality Problems in India
Source: Srikanth, 2009

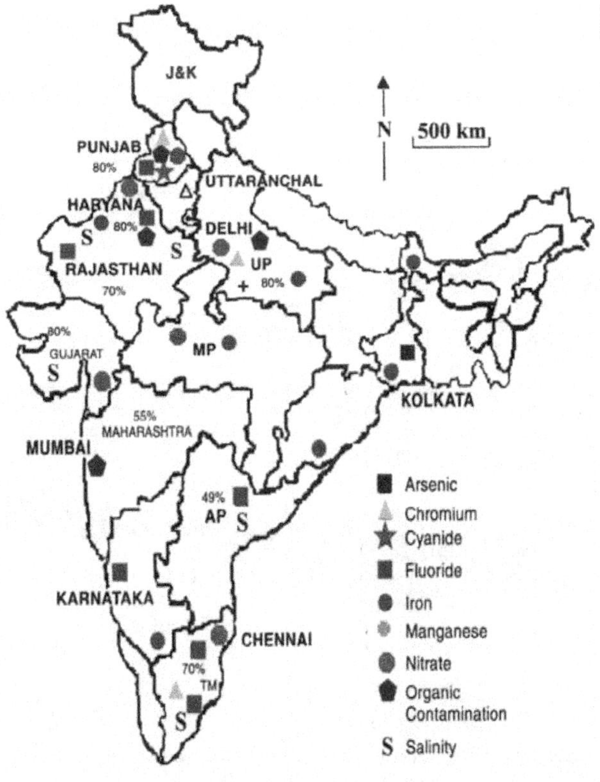

The problem of high fluoride concentration in groundwater resources is one of the most important toxicological and geo-environmental issues in India and about million people in villages are consuming water having high fluoride (Agrawal et al., 1997; Choubisa, 1998; Susheela, 1999). In India large known deposits of fluoride are located in Amba-Dongar and Karipani and a small occurrence of fluorspar in Ajmer district are reported (GSI, 1963) in Khairot and Barla. Plants growing in the vicinity of industries such as aluminum smelting can have substantially increased fluoride contents (Adriano, 1986).

Where intensive agriculture is practiced, nitrate levels in ground water are high. Farmers have a tendency to apply fertilizers at levels somewhat more than that recommended by the scientists or that required by the crop. Of the three nutrients Nitrogen, Phosphorus and Potassium, Nitrogen is most amenable to leaching. The leaching of fertilizers and pesticides into groundwater is influenced by a number of factors such as soil characteristics, levels of use, the timings of application, depth to water table, irrigation practices, nature of crops cultivated etc. In India documentation on the incidence of groundwater contamination in general and use of agrochemicals in particular is limited. Studies by Singh (1975), Singh and Sekhon (1976), Singh et al. (1987) concluded that agrochemicals pollute groundwater. A Government of India evaluation survey in 1999 noted that 142,000 habitations consume water that has excessive quantities of fluoride, iron, nitrate, arsenic and salinity. In Jharkhand and Chhattisgarh, uranium pollution is now coming on to the horizon as a water quality problem as has mercury pollution in West Bengal. In many cases, these sources of water have been rendered unsafe for human consumption as well as for other activities such as irrigation and industrial needs.

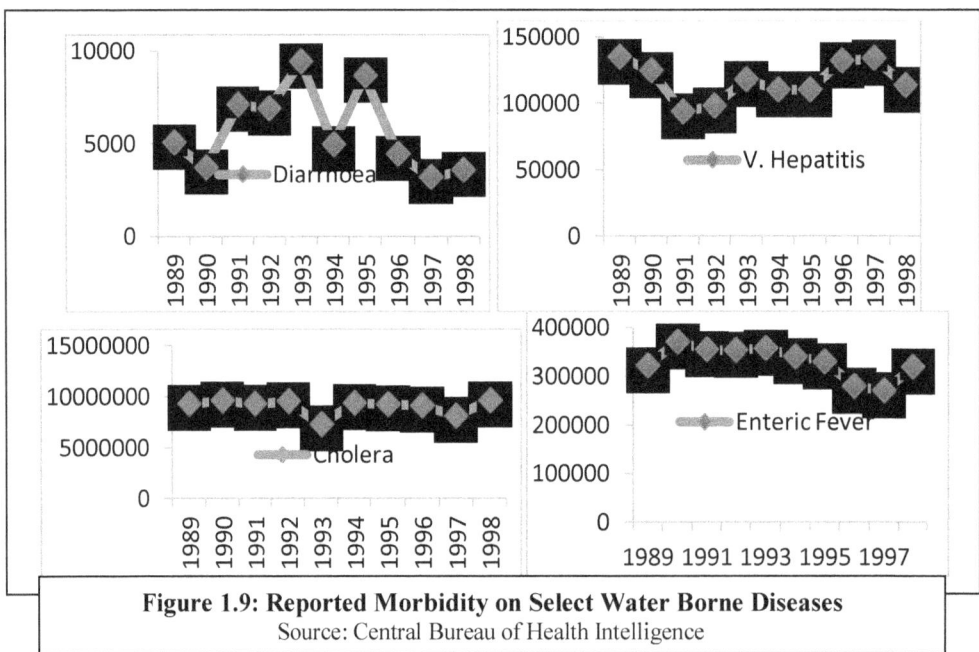

Figure 1.9: Reported Morbidity on Select Water Borne Diseases
Source: Central Bureau of Health Intelligence

Water quality is now being recognized in India as a major crisis. Water pollution is a serious problem as almost 70% of India's surface water resources and a growing number of its groundwater reserves have been contaminated by biological, organic and inorganic pollutants (Rao and Mamatha, 2004). *All of India's 14 major river systems are heavily polluted*, mostly from the 50 million cubic meters of untreated sewage discharged into them each year. The domestic sector is responsible for the majority of the wastewater generation in India (Rao and Mamatha, 2004). There are many places in India where groundwater is questionable or unsatisfactory. Millions of people in rural areas and low-income semi-urban communities are dependent on groundwater. These populations are most vulnerable to waterborne diseases as there is evidence of contaminated groundwater leading to outbreak of endemic diseases (WHO, 2004). So, much of this disease burden is found at places where the use of untreated water from shallow groundwater sources is common in both rural and semi-urban settlements (Pedley and Howard, 1997). The basic question in the production of drinking water is how to get rid of potentially dangerous chemicals and microorganisms without introducing new hazards and problems which might pose new and different threats to human health. Bio-films are formed in distribution system pipelines when microbial cells attach to pipe surfaces and multiply to form a film or slime layer on the pipe. Probably within seconds of entering the water distribution system, large particles, including microorganisms, absorb to the clean pipe surface (Clark et al., 2004). Factors that affect bacterial growth on bio-films include water temperature, type of disinfectant and residual concentration, assimilable organic carbon level, biodegradable organic carbon level, degree of pipe corrosion, and treatment/distribution system characteristics (Hunter *et al.*, 2001). Water treatment and distribution system, if not properly operated and maintained, can be a source of disease outbreak affecting large population. It is likely that problems of water scarcity, mismanagement, and water-related disasters will intensify due to increasing population, the rising demand for water for agriculture and other uses, and greater climatic variability or climatic change (Namara et al., 2010). Therefore, the monitoring and surveillance of quality of raw water sources and treated water need to be enhanced (Hamzah et al., 1997).

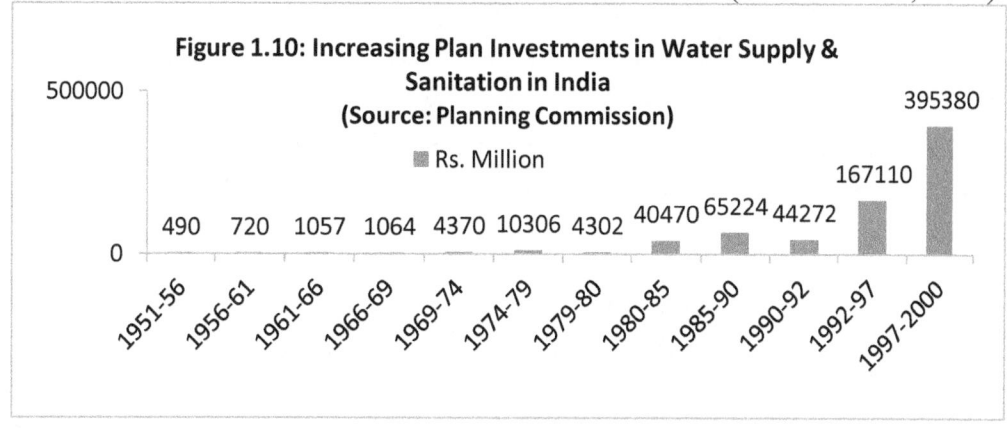

Figure 1.10: Increasing Plan Investments in Water Supply & Sanitation in India (Source: Planning Commission)

1.6 STATUS OF PUNJAB

A land of five rivers- Indus, Ravi, Beas, Sutlej and Ghaggar, Punjab was left with three perennial rivers- Ravi, Beas and Sutlej and one seasonal river Ghaggar, after partition in 1947. Besides this, three internationally important wetlands, two nationally important wetlands, several state wetlands, canals and drains, ponds and reservoirs exist in the state. Punjab being a very small state has very narrow range of both latitudinal and longitudinal extent which are to the tune of $29^{o}33'$ N - $32^{o}31'$ N and $73^{o}55'$ E - $76^{o}55'$ E, respectively (Mathauda et al., 2000). The state of Punjab is located in the north-west of India, bordered by the states of Jammu & Kashmir (north), Himachal Pradesh (east), Haryana and Rajasthan to the south. Pakistan lies to the west of Punjab.

Figure 1.11: STATUS MAP OF PUNJAB

Source: Punjab Remote Sensing Centre, Ludhiana

Out of 243.59 lacs of people in Punjab, 160.96 lacs reside in 27.96 lacs rural households and rural literacy rate is 71% males and 57.70% females (PRWSS, 2006). The state of Punjab is a part of the Indo-Gangetic plains formed by the alluvial deposits by rivers. Geo-morphologically the area has been divided into eight geomorphic units *viz.* hills, table land, intermountain valley, piedmont plains, alluvial plains, sand dunes, palaeo-channels and flood plains. As a consequence of diversity in the natural environment like climate, topography, parent rocks, drainage and vegetation cover spread over a span of time, the soils of Punjab developed largely on alluvium, vary widely and show difference in their nature, properties and profile development. In terms of soil type, the state can be divided into three zones –

a) **South-Western Punjab**: This region covers the tehsils of Fazilka, Muktsar, Bathinda, Mansa and parts of Ferozepur, which border Haryana and Rajasthan states in the southwest. The soil is predominantly calcareous, developed under hot and arid to semi-arid conditions. The pH value ranges from 7.8 to 8.5, which shows that the soil is normal in reaction. Grey and red desert, calsisol, regosol and alluvial soils are found in this zone. The soil of southwestern Punjab can further be sub-divided into two categories (i) Desert Soil and (ii) Sierozem Soil.

b) **Central Punjab**: The soil of this zone has developed under semi-arid condition. The soil is sandy loam to clayey with normal reaction (pH from 7.8 to 8.5). The soil covers the districts of Sangrur, Patiala, Ludhiana, Jalandhar, Kapurthala, Amristar, parts of Gurdaspur, Ferozepur and fringes of Kharar tehsil of Rupnagar district. Problem of alkalinity and Salinity is quite acute, especially in districts of Amristar, Sangrur, Ferozepur, Gurdaspur and Patiala. The soil of the central zone generally recognized as alluvial, falls into two categories, (i) Arid and Brown Soil and (ii) Tropical Arid Brown.

c) **Eastern Punjab**: The soil has developed in the sub-humid foothill areas bordering Himachal Pradesh covering eastern parts of Gurdaspur, Hoshiarpur, Rupnagar and northeastern fringes of Patiala district. Because of the undulating topography and fair amount of rainfall, normal erosion is quite common. The fertility of the soil is medium to low and the texture is loamy to clayey. Two soil types are recognized in this zone (i) Grey Brown Podzolic Soil (ii) Reddish Chestnut Soil.

Punjab is primarily an agrarian state and agriculture occupies the most prominent place in Punjab's economy. *As against an all-India average of 51 per cent, Punjab has 85 per cent of its area under cultivation.* Irrigation sector is the major user of water resources and accounts for about 85% of water consumption in the state (Source: Irrigation department, Punjab).

Table 1.1: Number of villages in Punjab in comparison to India having irrigation facility per 1000 villages, and their distribution by type of such facility in India (July-December 2002)			
State	Number of Villages having Irrigation Facility per 1000 Villages	Percent of Villages having Tube-well Irrigation	Percent of Villages having Other well Irrigation
Punjab	976	92.2	2.4
India	762	63.1	21.3

Source: India, Ministry of Water Resources 2007

Table 1.2: Groundwater sources in Punjab in comparison to India as on March31, 2003			
State	Available Groundwater Resources BCM/yr	Net Draft BCM/yr	Level of Groundwater Development [%]
Punjab	16.79	16.40	97.66
India	360.80	149.97	41.57

Source: India, Ministry of Water Resources 2007

After the onset of green revolution, the irrigation demand started increasing sharply due to cultivation of irrigation intensive crops of wheat and paddy in mid 1970s.

Punjab, the most stunning example of green revolution in India, is now at the crossroads. The dominance of rice and wheat monoculture cropping pattern over the years has led to overexploitation of ground water resulting in rapid decline of water table in the entire state. Besides over intensification of agriculture over the years has led to reduced soil fertility and micronutrient deficiency, non-judicious use of farm chemicals & problems of pesticide residue, reduced genetic diversity, soil erosion, atmospheric and water pollution and overall degradation of the rather fragile agro ecosystem of the state. *The state surface water resources are being fully utilized in sustaining the intensive agriculture practices. The groundwater is being overexploited to meet the increasing demands of water for irrigation intensive agricultural practices.*

Figure 1.12: Irrigation by Surface/Groundwater sources in Punjab
Source: Statistical Abstract of Punjab, 2005

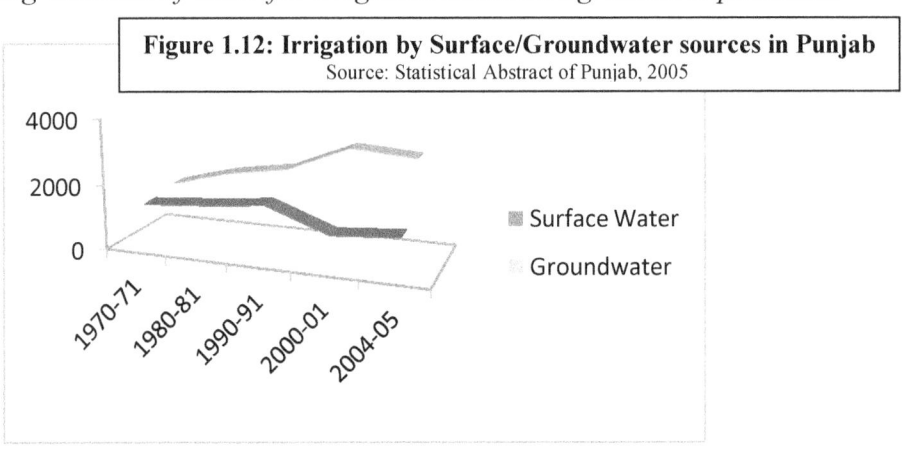

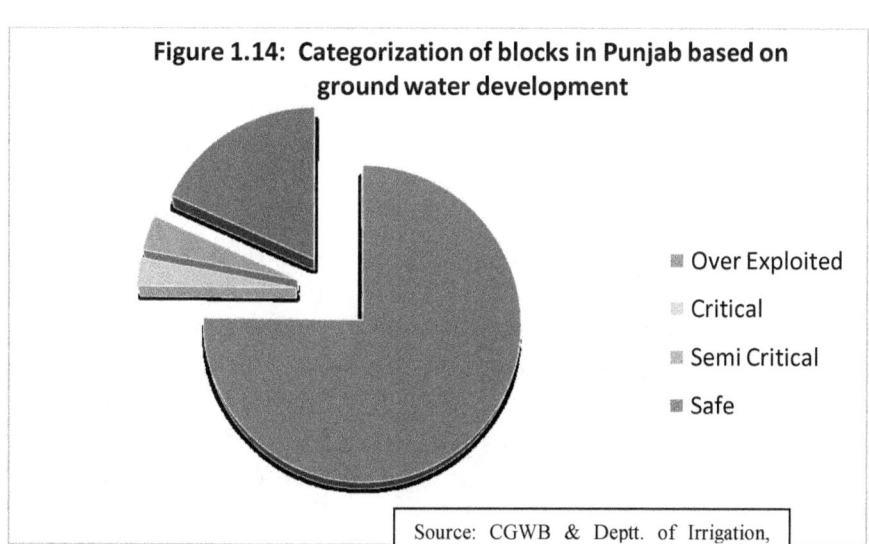

Legend

— MAIN CANAL
···· BRANCH CANAL
···· DISTRIBUTORY CANAL

Figure 1.14: Categorization of blocks in Punjab based on ground water development

- Over Exploited
- Critical
- Semi Critical
- Safe

Source: CGWB & Deptt. of Irrigation,

Groundwater Resources Estimation Committee (GEC, 2004) estimated the present groundwater development (ratio of gross ground water draft for all uses to net ground water availability) in the state to be 145%. Out of 138 blocks of the state, all the blocks of various districts like Amritsar (16), Bathinda (4), Faridkot (2), Jalandhar (10), Moga (5), Kapurthala (5), Sangrur (12), Fatehgarh Sahib (5), Patiala (8), Mansa (5), Nawan Shahr (3) and Ludhiana (10 out of 11) Ferozepur (7 out of 8), Gurdaspur (6 out of 10), Rup Nagar (2 out of 3), Hoshiarpur (2 out of 4) have been found to be overexploited leading to sharp depletion of the water table in these districts (CGWB, 2006). It is predicted that in about 66 percent area of the central districts the depth of water table would recede to 50 m by the year 2030 (Bhatt, 2010; Hira *et al.,* 2004; 2002).

Table 1.3: Falling Water Levels
(The districts are included in this list if they experience a drop in groundwater level of over 4 meters for 20 year period or a drop of over 2 meters over a ten year period.)

Period	Districts of Punjab
As on 1982-2001	Amritsar, Bathinda, Ferozepur, Jalandhar, Kapurthala, Ludhiana, Mansa, Moga, Patiala, Ropar, Sangrur.
Added in 1983-2002	Faridkot, Fatehgarh.
Removed in 1994-2003	Faridkot, Jalandhar, Kapurthala, Ropar.
Added in 1995-2004	Faridkot, Jalandhar, Kapurthala, Ropar, Gurdaspur, Hoshiarpur, Nawan Shahr.

Source: Centre for water Policy, 2005

Figure 1.15: Map - Categorization of blocks in Punjab based on groundwater development, 1999

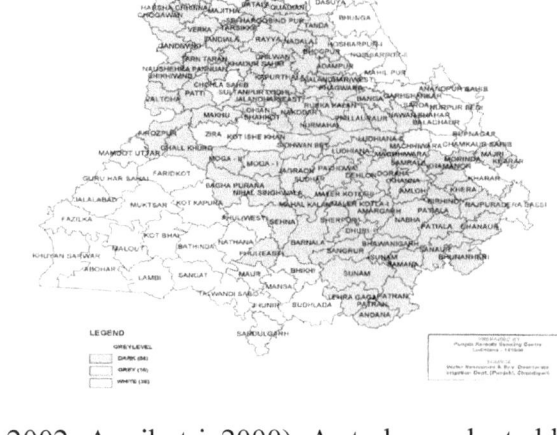

The policy of subsidy on fertilizers encourages the farmers for excessive use of chemical fertilizers, which adversely affected soil and water quality over time. It has increased more than eight times in the past 35 years from 213 thousand tons in 1970-71 to 1694 thousand tons in 2005-06. Use of pesticides in Punjab has increased from 3200 Metric tons in 1980-81 to 7300 in 1994-95 but came down to 5970 in the year 2005-06. Currently, the state consumes about 17 percent of total pesticides used in India (Singh, 2002; Agnihotri, 2000). A study conducted by Joia et al. (1976), on wheat flour samples collected from Jalandhar, Patiala, Sangrur, Ludhiana, Faridkot,

Amritsar and Chandigarh found residues of DDT, Benzene Hexachloride (BHC) and Hexachloro cyclohexane (HCH). Battu et al. (1978) found pesticidal contamination in vegetable oils and oil seed cakes in the samples collected from Ludhiana, Muktasar, Ferozepur, Sangrur and Khanna. Kalra and Chawla (1980) found DDT and BHC residues in the milk samples of the lactating women residing in Punjab. Chattopadhyay (1998) and Mathur et al. (2005) detected pesticides residue in the blood samples obtained from Patiala region and villages of Punjab, respectively. Results of the studies conducted by Punjab Pollution Control Board since 2002 confirmed the presence of pesticide residues in water and sediments of river Sutlej, Beas and Ghagger.

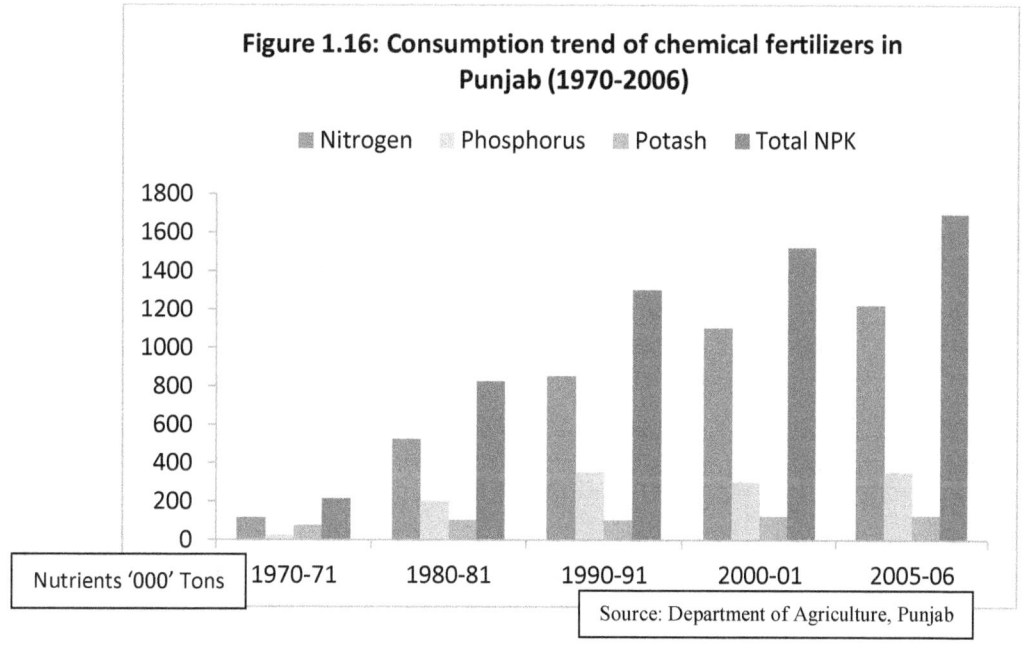

Figure 1.16: Consumption trend of chemical fertilizers in Punjab (1970-2006)

Nutrients '000' Tons

Source: Department of Agriculture, Punjab

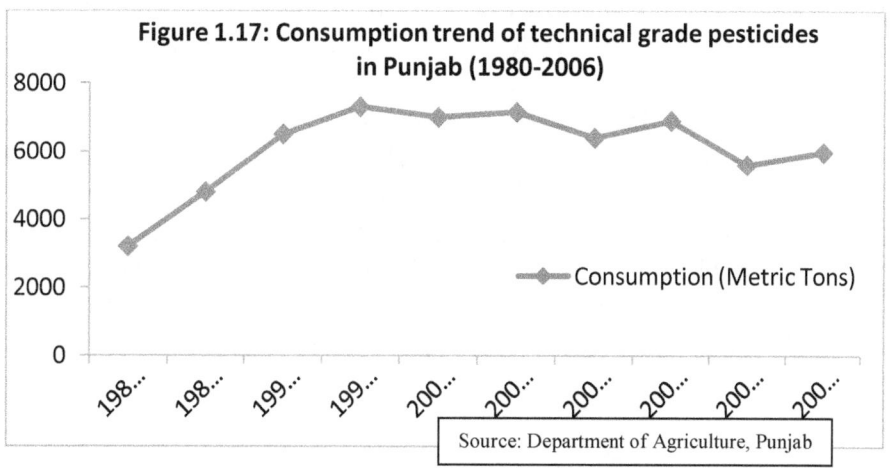

Figure 1.17: Consumption trend of technical grade pesticides in Punjab (1980-2006)

Source: Department of Agriculture, Punjab

The soil pollution load further increases due to their reaction products and residues. Some of the reaction products such as nitrates and phosphates find their way to surface waters and aquifers. The enrichment of surface waters with these nutrients by runoff from agriculture fields and by leaching causes eutrophication. A study conducted by Bajwa *et al.* (1993), indicated that groundwater samples collected from 21 to 38 m deep tube-wells located in cultivated fields of Dehlon, Sudhar, Ludhiana, Kartarpur, Jandialaguru and Malerkotla blocks of Punjab contained higher nitrate concentration; in areas specially under rice, maize and orchards. For only 1.5% landmass of the country, Punjab consumes about 17% of pesticides used in India. Out of these, >90% of the pesticide are being used in cultivation of cotton, rice and vegetables (Singh, 2002). This is causing degradation of soil including nutrient imbalance, depletion of underground water table, abuse of pesticides and fertilizers leading to several environmental and health hazards.

Table 1.4: Nitrate, Fluoride and Heavy Metals/Arsenic affected Districts

Districts affected by excess nitrates (over 45 mg/L)	Districts affected by excess fluoride (over 1.5 mg/L)	Districts affected by toxicity in groundwater due to heavy metals/ arsenic
Bathinda, Faridkot, Ferozepur, Patiala, Sangrur	Ferozepur, Faridkot, Muktasar, Patiala	Ludhiana, Mandi Gobindgarh, Fatehgarh Sahib

Source: Centre for water Policy, 2005

Punjab, which has done remarkably well in the field of agriculture, is also on its way to rapid industrialization through development of Small, Medium and Large scale industries. A notable feature of the industrial scenario of the Punjab is its small-sized industrial units. There are 194,000 small scale industrial units in the state in addition to 586 large and medium units. Industrial units in the state are namely divided into three:

- **Agro-based industrial** units include Food Products and Beverages, Vanaspati Ghee, Dairy, Textile, Hosiery and Garments, Sports, Leather, Wood Products, Paper, Recorded Media, Paper Industries, etc.
- **Machinery** units include Fabricated Metal, Machinery and Equipment, Office Machinery, Electrical Machinery, Communication Equipments, Medical Instruments, Motor Vehicle Parts, Cycle and Parts, etc.
- **Chemical** units include Chemicals, Rubber and Plastic, Metallic, Non-metallic, etc.

Table 1.5: Internationally Renowned Indian Companies Working in Punjab

Company	Product
Ranbaxy	Medicines
Hero Cycles, Avon Cycles	Cycles
Punjab Tractor Ltd.	Swaraj Tractors and Combine Harvester
Oswal Woolen Mills	Monte Carlo , Casablanca
Oswal Knit India Ltd.	Pringle
JCT Textiles, DCM	CTV Picture Tube , Steel rope ,Castings
Birla_VXL(OCM)	Woolen fabric
JIL	Maltova , Viva , range of wines and liquor
Gujrat Ambuja	Cement
Godrej	Washing Machine
ACC	Cement
SIEL	Chemicals, Vanaspati
Abhishek	Denim Fabric

(Source: Department of Industries and Commerce, Punjab)

Punjab leads in manufacture of machines and hand tools, printing and paper machinery, auto parts and electrical switchgears. Punjab produces around 75% of bicycle and bicycle parts, sewing machines, Woolen and other Hosiery items, Shoddy blanket and jacket clothes and sports goods.

The rapid industrialization, urbanization and non sustainable development have led to the establishment of number of industries/ commercial complexes which in turn has aggravated the cause of water pollution in the State which now has become graver during the past few years. The Waste/wastewater generated from the urban areas, industries and different commercial activities is discharged into rivers, streams and drains or dumped on the land, resulting into pollution. *The pollution potential in the state has increased, which has not only degraded the quality of water used for irrigation as well as for drinking purposes, but also has affected the flora & fauna.* The wastewater generated from industries, various households & animal sheds and effluent from septic tanks overflows into the open surface drains and ultimately finds its way into the soil or ponds, thus making them highly polluted (PRWSS, 2006). Polluted and clogged open drains lead to poor sanitation and unhygienic conditions.

Table 1.6: Punjab – Rural Drinking Water Quality Scenario										
		Villages with Various Scarcity Parameters							Villages having scarcity due to Multiple parameter	Villages above acceptable limit but below cause of rejection
Total inhabited villages	Total Number of scarcity villages	Depth >15m, Distance >1.6km	Excess iron	DS	Excess Fluoride	Total hardness	Chloride	Bacteriological problem		
12402	11849	2753	967	1771	851	1220	283	65	389	3550
(Source: PWD, Public Health Branch, 2004, Govt. of Punjab)										

Data from the National Sample Survey Organization (1999) indicates that *82.7% people in rural areas of Punjab use tube-well/hand-pumps as their principal source of drinking water* as it is considered that Groundwater is a low cost high quality water source whereas surface water is a high cost low quality source. But the Groundwater quality depends on depth of water table, type of geological formations, amount of water recharging the aquifer, source of pollution etc. The communities are exposed to high risk of infection due to water- and excreta-related diseases like cholera, hepatitis, malaria, diarrhea etc. Health risk also arises where villagers are using shallow hand pumps located near the ponds or soak pits for drinking water. Water that receives runoff from land used for agriculture and livestock farming is likely to have pesticides, fecal matter, and other constituents that were applied to the surrounding grounds. The land application of poultry and dairy wastes containing organic forms of nutrients may reach nearby rivers and streams during storm events. Increases in discharge are caused by storm-water runoff, which is associated with nonpoint overland flow from urban and agricultural land uses (Frick et al., 1996). Livestock and poultry operations have properties that can generate large amounts of manure and waste. When improperly managed, nutrients, bacteria, and other pollutants can contaminate drinking water supply. Ellis (1991) suggests relocating cattle and other livestock from the vicinity of drinking water sources. In 2004, nearly 4,75,000 out-patient cases were reported in Punjab for diseases caused due to water- and sanitation-related conditions. Approximately 173 deaths were reported due to these diseases (*Health and Family Welfare, Punjab, 2004*). *There is a problem of unhygienic disposal of human faeces in the absence of sanitation facilities*. Because of the presence of pit latrines close to the water sources, lack or little environmental protection, and poor catchments management, there may increased contamination of the sources of water (Zamxaka et al., 2004). Consequently, there is widespread practice of open defecation in rural areas. Contamination by sewage or human excrement presents the greatest danger to public health associated with drinking water, and bacteriological testing continues to provide the most sensitive means for the detection of such pollution (Hrudey et al., 2003). Groundwater can be and often has been contaminated by people's aboveground activities.

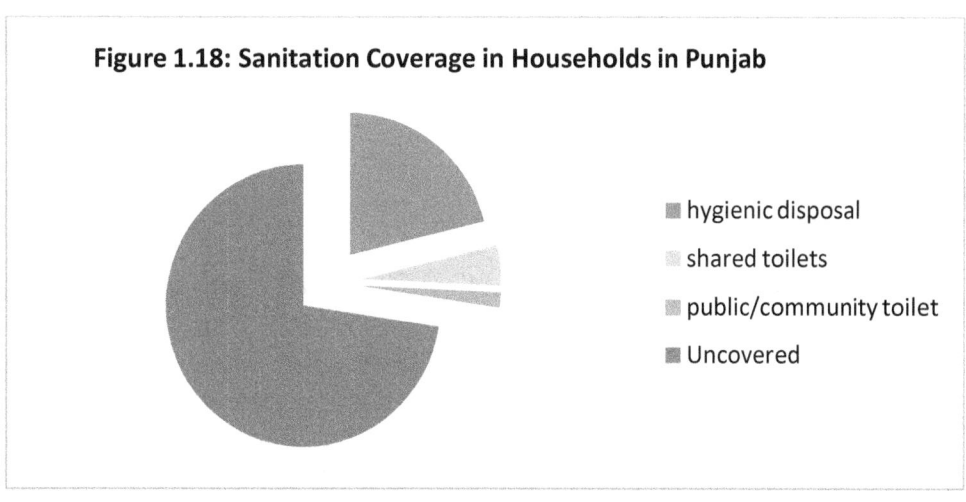

Figure 1.18: Sanitation Coverage in Households in Punjab

- hygienic disposal
- shared toilets
- public/community toilet
- Uncovered

(Source: Department of Water Supply and Sanitation, Punjab, 2006)

A number of Indian states including Punjab have commenced rural water supply and sanitation projects on the reform guidelines enunciated in Government of India's Swajaldhara policy and Total Sanitation Campaign. In Punjab, drinking water sources include organized piped water supply schemes as well as a number of private spot sources. Properly treated water may become contaminated by microorganisms (bacteria, protozoa etc) again after it leaves the treatment plant and enters the distribution system. Factors that affect bacterial growth on bio-films include water temperature, type of disinfectant and residual concentration, assimilable organic carbon level, biodegradable organic carbon level, degree of pipe corrosion, and treatment/distribution system characteristics (Vlek *et al.*, 2006).

The table presents the findings of Rajiv Gandhi Drinking Water Mission as well as progress of rural water supply until March, 2006.

Table 1.7: Coverage of Rural Habitations with Water Supply in Punjab (Mar-2006)							
Type of Habitations	Fully covered		Partially covered		Not covered		Total
	Nos.	%	Nos.	%	Nos.	%	Nos.
Main habitations	6319	52	2861	23	3087	25	12267
Other habitations	845	36	731	31	762	33	2338
Total habitations	7164	49	3592	25	3849	26	14605
(Source: Department of Water Supply and Sanitation, Punjab, 2006)							

Table 1.8: Breakup of villages with Quantity of Water Supply			
Less than 40 lpcd or not having provisions for livestock	55 lpcd	70 lpcd	100 lpcd
6864	34	1704	14
(Source: Department of Water Supply and Sanitation, Punjab, 2006)			

Water has always been a contentious and tricky affair in India due to socio-economic-political and ecological reasons. Factors like caste-class differences, heterogeneity of farmers, rural–urban dichotomy, and extreme different ecological conditions have influenced the water management. To complicate further, vote bank politics, lack of coordination between irrigation bureaucracy, policy making and various sectoral departments carrying out their own water programmes, have affected water management in a diverse manner to people. *Understanding how these different policies and programs influence water and its resources at the community level is one of the unexplored issues.* Most of the data on groundwater quality is unfortunately limited to shallow aquifers whereas Total Dissolved Solids (TDS), Fluoride, Iron, Hardness, Nitrates and Chloride are the major water quality parameters relevant to domestic needs in Punjab (Punjab Rural Water Supply and Sanitation Project, 2006). Previously, attempts have been made to assess the underground water quality in different districts (Rupnagar, Sangrur, Muktsar, Mansa, Fatehgarh Sahib) of Punjab (Minakshi et al., 2006; 2004; Singh and Bishnoi, 2004; 1993; Patel et al., 2001; Brar and Chibba, 1997), but such comprehensive information about other districts of Punjab is also required (Minakshi et al., 2006).

1.7   AIMS AND OBJECTIVES:

❖ To monitor the physicochemical and microbiological quality of rural drinking water supplies to the rural areas of district Patiala.

❖ To compare the physicochemical parameters and microbiological quality of drinking water with respect to International and Indian Standards.

❖ To give emphasize for the awareness in common people about water quality.

❖ To explore and compare the water quality of different blocks of Patiala district.

❖ To study the groundwater quality and its mapping

❖ To study the groundwater quality with respect to Temperature, pH, Electrical Conductivity, Total Dissolved Solids, Total Suspended Solids, Total Hardness, Calcium, Magnesium, Total alkalinity, Chloride, Fluoride, Nitrate and Total Coliforms.

❖ To examine feasibility of different classification criteria

❖ To assess the effect of various quality parameters on human health

❖ To analyze the utility of water for domestic purposes

1.8   OVERVIEW OF THE BOOK

To facilitate the study, the Book is spread out in the chapters as given below:

Chapter 1   Introduction of the problem and objectives of the study;

Chapter 2   Contains Review of the Literature;

Chapter 3   Discuss about Patiala District from where the samples of water are monitored;

Chapter 4   Discuss the drinking water quality parameters;

Chapter 5   Include the drinking water quality analysis of the rural areas of district Patiala;

Chapter 6   Includes the result and discussion of the problem;

Chapter 7   discuss the conclusion and recommendations of the study; and

Chapter 8   includes summary followed by
Bibliography at the very end of the book.

1.9   RELEVANCE AND NEED OF PRESENT STUDY

In Patiala the surface water/ groundwater is polluted by agriculture, industrial waste discharged by industries and by municipal waste. These discharges contain organic load, heavy metals and pesticides and fertilizers. There are about 150 medium and large industrial units working in Patiala. Chemical quality of ground water occurring at shallow depths as well as at deeper depths has been evaluated in this industrialized city. The present research work may have the following significance:

➢ The assessment of drinking water quality and contamination can be useful for decision makers and daily human life, well being.

➢ Output of research can be informative for urban and rural water planning.

➢ "Save Water" and "Safe Water", slogans can be considered and implemented seriously.

➢ The need of the time, Sectoral Water Management and Integrated Water Management can be introduced and implemented.

➢ This research can be the basis of future research of drinking water quality and contamination using different spatial and temporal datasets.

1.10   LIMITATIONS OF THE STUDY

In this study, the following limitations were faced:

- For effective maintenance of water quality through appropriate control measures, continuous monitoring of large number of quality parameters is essential. However it is very difficult and laborious task for regular monitoring of all the parameters even if adequate manpower and laboratory facilities are available. The limited number of water quality parameters studied may bring out the limited reality of water.
- Convenient non-probability sampling method was employed to collect water samples which may not indicate the water quality of the total block or the district.
- Time constraint, by which any book is bound, may not be able to map the exact picture.

Chapter-2

Contaminated drinking water poses a major health threat to human beings worldwide. In recent years, due to pollution and change in climatic conditions the rivers, canals, tube-wells and hand-pumps which provide water, may not be suitable for drinking purposes. Numerous severe cases of groundwater contamination have been documented worldwide (Farooq et al., 2010; Mor et al., 2008; Shivkumar and Biksham, 1995). Study of quality of drinking water and establishing standards is at least 4000 years old (Raucher, 1996). Quality assurance is a set of operating principles that if strictly followed during sample collection and analysis will produce data of known and defensible quality. That is the accuracy of results can be stated with a high level of confidence (APHA, 1995). Huang and Xia (2001) emphasized the in-depth research of the related barriers and the relevant mitigation approaches to enhance sustainability of water quality management systems. Various studies have been carried out in India including Punjab and abroad regarding the quality of water and its suitability for plants, animals and human beings.

pH and Alkalinity is the important physicochemical water quality. In water, a small number of water (H_2O) molecules dissociate and form hydrogen (H^+) and hydroxyl (OH^-) ions. If the relative proportion of the hydrogen ions is greater than the hydroxyl ions, then the water is defined as being acidic. If the hydroxyl ions dominate, then the water is defined as being alkaline. The relative proportion of hydrogen and hydroxyl ions is measured on a negative logarithmic scale from 1 (acidic) to 14 (alkaline): 7 being neutral (Friedl *et al.,* 2004). Acidic or Alkaline water affects mucous membrane in the mouth and put a lot of troubles in stomach. Salinity, another physicochemical water quality, is measured by comparing the dissolved solids in a water sample with a standardized solution. The dissolved solids can be estimated using total dissolved solids or by measuring the specific conductance. Specific conductance, or conductivity, measures how well the water conducts an electrical current, a property that is proportional to the concentration of ions in solution. Conductivity is often used as a surrogate of salinity measurements and is considerably higher in saline systems than in non-saline systems (Dodds, 2002). Turbidity and Suspended Solids are the other important water quality parameters. Turbidity refers to water clarity. The greater is the amount of suspended solids in water, the murkier it appears, and the higher the measured turbidity. The major source of turbidity in the water is phytoplankton, clays and silts, re-suspended bottom sediments, and organic detritus from stream and/or water discharges. The source of these sediments includes natural and anthropogenic activities, such as natural or excessive soil erosion from agriculture, forestry or construction, urban runoff, industrial effluents, or excess phytoplankton growth (USEPA, 1997). Municipal, agricultural, and industrial discharges can contribute ions to receiving waters or can contain substances that are poor conductors like organic compounds, changing the conductivity of the receiving waters. Thus, specific conductance can also be used to detect pollution sources (Stoddard et al., 1999). The most important pollutants are probably chloride and nitrate, which can be derived from leaking sewers, landfills, un-sewered sanitation, livestock farming and agricultural fertilizers. High chloride concentrations may also indicate saline intrusion in coastal areas or the use of salt for road de-icing in cold climates. These pollutants can be indicators of impact from both rural and urban activities. Nitrate in particular can be problematic, as un-sewered sanitation and agriculture often occur in close proximity. Since these parameters cannot be removed or eliminated by means of water disinfection and can bring about a lot of health hazards, a complementary unit must be added to water treatment processes or these resources should be considered as undrinkable water resources (Chapman, 1996).

Poor water quality can be the result of natural processes but is more often associated with human activities and is closely linked to Agricultural and industrial developments. Intensification of agriculture, increasing productivity, rapid extension of irrigation, fertilizer application and pest control, put unanticipated adverse impacts on the quality of underlying groundwater. Microbiological health risks remain associated with many aspects of water use, including drinking water in developing countries, irrigation reuse of treated wastewater and recreational water contact (Grabow, 1991). The information on the quality of groundwater yields valuable knowledge regarding their possible effects on physicochemical properties of the soil and its productivity (Sharma and Minhas, 2004). Climate change, the evolution of new waterborne pathogens, and the development and use of new chemicals for industrial, agricultural, household, medical, and personal use have raised concern as they have the potential to alter both the availability and the quality of water (Kolpin et al., 2002). All of these activities have costs in terms of water quality and the health and integrity of aquatic ecosystems (Meybeck, 2004).

Human activities such as mining and heavy industry can result in higher concentrations of trace metals than those that would be found naturally (Charlet and Polya, 2006). Elevated concentrations of trace metals can have negative consequences for both wildlife and humans. Trace metals can affect function of pH, oxidation-reduction state, and organic matter content of the water. Besides, an important environmental problem (Rayms-Keller et al., 1998), heavy metals constitute some of the most hazardous substances that can bio-accumulate (Tarifeno-Silva et al., 1982). Bioaccumulation is a process in which a chemical pollutant enters into the body of an organism and is not excreted, but rather collected in the organism's tissues (Zwieg *et al.,* 1999). Metals that are deposited in the aquatic environment may accumulate in the food chain and cause ecological damage while also posing a threat to human health (Ermosele *et al.,* 1995). Pollution by the heavy metals also needs to be cared as it is anthropogenic impacts like industrial discharges, domestic sewage, nonpoint source runoff and atmospheric precipitation which are the main sources of toxic heavy metals that enter aquatic systems (Langston *et al.,* 1999). However, metals also occur in small amounts naturally and enter aquatic systems through ore-bearing rocks, windblown dust, forest fires and vegetation.

Holt et al. (2000) was of the view that surface water can be contaminated through direct or indirect emissions and ground water can be contaminated by leaching from the soil. There are various ways by which pollutants penetrate into ground water; inadequate storage of harmful substances, refuge dumps, transport accidents, infiltration of polluted rain water (Ashour and Hung, 2000), fertilizers etc. (Mattikalli and Rechards, 1996). Groundwater control is very different from the control of surface water bodies. It implies accurate information on the distance between aquifer and the surface of ground, on the volume of the ground water body, on the nature of its underlying stratum, on the velocity of flow and on the chemical characteristics of the various water bearing formations (Koppe, 1973). The diversity and number of existing and potential source of chemical contamination are very large. It is estimated that there are between ninety thousand and one hundred thousand chemicals in regular use but that as few as three thousand accounts for about 90% of the total mass used (Holt et al., 2000). Funari and Ottaviani (1997) presented some of the main aspects of the risk to human health associated with the possible exposure, through drinking water, to carcinogenic and non-carcinogenic chemical substances and microbiological agents. More research is needed to assess the relationship between drinking water chemistry and human health.

Harrison et al. (2000) compared the drinking water supplies in the United Kingdom and found that private supplies follow less stringent sampling and testing regimes than public supplies. Information regarding the private drinking water supplies is desperate and poorly defined. Majority of breaches of standards were due to increased concentrations of nitrate in 270, magnesium in 21, manganese in 17, iron in 15 and turbidity in 15 cases. Regular sampling of private drinking water supplies remains necessary to prevent risk to health from a wide variety of toxic contamination.

Hacioglu and Dulger (2009) studied the water samples collected from Biga Stream, Turkey and concluded that there is a great potential risk of infection of waters as the water quality was very low.

Mallo et al. (2001) studied 100 water samples in Tandil region of Argentina for pH, temperature, hardness, chloride, calcium, sulphates and nitrates and grouped them in 5 classes based on difference in levels of parameters using statistical methods; after concluded that drinking waters and subterranean water from the regions of Argentina were suitable for human consumption whereas surface water is highly polluted.

According to Karbassi et al. (2008); Najafpour et al. (2008) and Singh et al. (2004), the anthropogenic discharges constitute a constant polluting source, whereas surface runoff is a seasonal phenomenon, largely affected by climate within the basin. Human activities are a major factor, determining the quality of the surface and groundwater through atmospheric pollution, effluent discharges, use of agricultural chemicals, eroded soils and land use (Niemi *et al.*, 1990). Run offs with substantial agricultural and other land use experience increased inputs and varying compositions of organic matter (Sickman *et al.*, 2007) and excessive concentrations of phosphorus and other nutrients from fertilizer application and other releases (Easton *et al.*, 2007).

The underground water in Karachi in Pakistan is also quite unsafe due to mixing of sanitary and sewage systems. About nine hundred out of eleven hundred water samples collected from different pipelines and hotels by Karachi Metropolitan Corporation Laboratory were found unfit for human consumption (SEGMITE, 1999).

Sadeghi et al. (2007) investigated the quality of rural drinking water supplies in Iran and Data showed that 30.2% of the villages under study had contaminated water resources.

Okafo et al. (2003) indicates urban and farm runoffs, discharges from sewage treatment facilities, failing septic systems, wildlife, farm animals and direct faecal contamination by humans and animals as the major sources of contamination.

Oxygen levels in water can be reduced through over fertilization, by run-off from farm fields and sewages containing phosphates and nitrates, as to stabilize nitrogenous wastes to ammonia then to nitrite and nitrate, bacteria consume high amount of oxygen.

Tehounwou et al. (1997) conducted studies to assess the physical, chemical and bacteriological qualities of drinking water Mbandjock, Cameroon. Their results indicated that vast majority of drinking water sources possess acceptable physical and chemical qualities as per WHO standards. However bacteriological analysis revealed that only the water treated by Cameroon National Water Company and Sugar Processing Company were acceptable for human consumption. All spring and well waters presented evidences of faecal contamination from human and/or animal origin.

Study by Crabill et al. (1999) on physical and chemical parameters (including stream flow, pH, temperature, dissolved ion concentrations and biological oxygen demand) indicated that the water of Oak Creek, Argentina was of high quality. However faecal coliform enumerations of Oak Creek demonstrated an annual deterioration of water quality during the summer season of 1994, 1995, 1996.

Edberg et al. (1997) stressed on the biological monitoring coupled with physicochemical monitoring to establish a long term history of the source.

Colour in natural water may be due to organic matter which originates from soil, peat and decaying vegetation or inorganic ions. Iron, copper and manganese present in ground and surface water may impart red, blue and black hue, respectively, which may be enhanced by bacteriological processes. Furthermore, colour producing organic substances can react with chlorine to produce undesirable levels of chlorination byproducts (WHO, 1996). Objectionable odours and flavours in drinking water are a constant concern to consumers and public water suppliers. The organoleptic properties of drinking water can be naturally induced or manmade. There are substantial grounds to support the possibility that unfamiliar drinking water odours might reveal the presence of substance, which pose a potential health risk (Jardine et al., 1997). Compounds in water that are perceived as giving it a taste, are generally inorganic substances present in concentrations higher than organic pollutants. The salt concentrations in water are approximately the same as in saliva for the water to taste neutral. Abundance of blue-green algae can cause drinking water taste and odour problems, in treatment plants during the summer months. Cyanobacterial toxins are produced by terrestrial, fresh, blackish and seawater cyanobacteria of cosmopolitan occurrence. These toxins bring acute and chronic hazards to human and animal health. Human health problems are associated with the ingestion of and contact with cyanobacterial blooms and their toxins (Szewzyk et al., 2000; Codd et al., 1997). According to Ling (2000) eutrophication due to excess of nitrogen and phosphorus in sources of drinking water leads to massive proliferation of cyanobacteria. The dominant species of cyanophyta can produce microcystins, a potent liver cancer promoter.

Boulay and Edwards (2001) reported the role of temperature and chlorine in copper corrosion byproduct released in soft water. Soft, low alkalinity drinking water tends to cause relatively high copper corrosion byproduct release in plumbing systems. Long term tests (6-8 months) in a synthetic, microbial stable soft tap water confirmed that lower pH and higher temperature increase release of copper in water.

Momba and Kaleni (2002) performed an experiment to test the drinking water, stored in polyethylene and galvanized steel containers by rural communities of South Africa and found that both types of containers support growth and survival of indicator bacteria. Bio-films in the drinking water distribution system can protect pathogens from disinfection and provide the inocula for periodic infestations (Smith et al., 2000).

Systematic analysis of fluoride research reveals that presence of low level of fluoride in drinking water does reduce dental caries, but surely adds to dental fluorosis. The latest estimates suggest that around 200 million people, from among 25 nations the world over, are under the dreadful fate of fluorosis. India and China, the two most populous countries of the world, are the worst affected (Ayoob and Gupta, 2006). Groundwater fluoride in high level has been reported (Agrawal et al., 1997; Maithani et al., 1998; and Datta et al., 1999) in all the 31 districts of Rajasthan with a serious health related issue in 23 districts. Several studies have indicated a possible link between arsenic and fluoride in drinking water (Wyatt et al., 1998). Their study showed a positive correlation. 52 districts out of 64 covering an area of $118.012 Km^2$ and approx. 40 million people show arsenic occurrences in groundwater in Bangladesh (Karim, 2000). The groundwater in 7 districts of West Bengal, India, covering an area of $37,000 Km^2$ with the population of 34 million, has been found contaminated with arsenic. Groundwater of Rajnandgaon district of Madhya Pradesh was also found contaminated with arsenic at 880 ppb (Chakkraborti et al., 1999).

According to Sisti et al. (1998) the effect of chlorine is markedly influenced by temperature. At a summer water temperature ($20^{\circ}C$) the efficacy of the chlorine concentration was found to be two to three times lower than that at winter temperature ($5^{\circ}C$). According to Sharma (2001) the ideal temperature of water for drinking purpose is 5-$12^{\circ}C$. Above $25^{\circ}C$, water is not recommended for drinking. The increase in temperature decreases palatability, because at elevated temperatures, carbon dioxide and some other volatile gases, which impact taste, are expelled. The chemical and biological activity of water, disinfection, coagulation and sedimentation processes are affected by temperature.

Nebbache et al. (2001) suggested that turbidity and nitrate concentrations peak during heavy rain episodes and are short term events. In terms of management it implies that water pollution caused by such events is also short time and can, therefore, be addressed at a local scale.

Betancourt et al. (2000) detected coliforms and high turbidity (6 NTU) in the finished water indicating poor operation of the filters and the subsequent interference with disinfection. Angjeli et al. (2000) investigated the drinking water cycle from its natural source to consumers, in suburbs of Albanian capital, by analyzing samples and verifying pollution levels in the microbiological and chemical setting. The most important pollution sources were found in the distribution network due to cross-contamination with sewers and illegal connections. The second pollution source was found around the extraction wells. Diarrheal diseases are major causes of morbidity and mortality among children in developing countries (Ansaruzzaman et al., 2000; Thevos et al., 2000). Diarrheal disease is one of the most important health problems related to water borne pathogens (Van Leeuwen, 2000). Germani et al. (1994) found water to be the potential source of contamination for diarrheal disease. Atef and Al-Kharabsheh (1999) monitored the Water quality of nine representative springs in the wadi Kufranja basin (Jordan). The chemical and biological contents of the spring water showed cyclic values due to water consumption and recharge during summer and winter. Electrical conductivity (547-1030 µS/cm), Nitrates (100 mg/L), faecal coliform bacteria (100 colonies/100ml) was found.

Saleh et al. (2001) reported that tap water from both Cairo and Giza, Egypt, was of higher quality than any of the bottled water with regard to analyzed anionic and cationic chemical constituents and within the permissible limits of WHO.

In a study of groundwater wells in agricultural southwestern Ontario (Canada), 35% of the wells tested positive for pesticides on at least one occasion (Lampman, 1995). Studies by Schottler et al. (1994) indicate that 55-80% of the pesticide runoff occurred in the month of June. The Netherlands National Institute of Public Health and Environmental Protection (RIVM, 1992) concluded that groundwater is threatened by pesticides in all European states. Wyatt et al. (1998) analyzed water samples taken from wells or storage tanks, direct sources for domestic supply in Northern Mexico, for the presence of lead, copper, cadmium, arsenic and mercury. 43% samples were found exceeding the limit for lead, 42% exceed the limit for mercury and 9% exceed the limit for arsenic.

Surface water and other drinking water sources have been polluted to different extents. The main pollutants present in drinking water sources are organic substances, ammonia, nitrogen, phenol, pesticides and pathogens. Some of these cannot be removed effectively by traditional water treatment processes like coagulation, sedimentation, filtration and chlorination, and the product water usually does not meet the national drinking water standards.

Oeziuerk and Yilmaz (2000) conducted a preliminary study on trace elements (arsenic, mercury, lead, cadmium, copper, chromium, iron, manganese and zinc) in drinking water. The obtained results showed that, in general, the trace elements concentrations did not increase WHO limits. Ryan et al. (2000) also proved that drinking water is a well recognized pathway of exposure to these metals. Another interesting health based work done by Vahter et al. (2002) aimed at reviewing exposure and health effects of cadmium, nickel, lead, mercury and arsenic manifested differently in women than men.

Somashekar et al. (2000) presented the date on the chemistry on forty eight tube wells' water collected from Channapatana town, India and surroundings. According to them quality of 80% of wells was unsuitable for drinking in terms of hardness, 50% in terms of magnesium and 20% in terms of nitrates and calcium.

Gupta *et al.* (1999) assessed the fluoride concentration and other parameters in 658 groundwater samples from villages in tehsil Kheragarh of Agra district and found 27% of the samples in range of 0.0 to 1.0 mg/L and 16% above 3.0 mg/L.

Prasad *et al.* (2008) studied the pH, electrical conductivity, TDS, TH, TA, calcium, magnesium, sodium, potassium, chloride, nitrate, sulphate and fluoride of 18 water samples collected from Lalsot urban area in Dausa district of Rajasthan and found the pH values in the study area range from 7.3 to 8.7 with the mean value of 7.93, Electrical conductivity from 402 to 2077 with a mean value of 1146.44 μ mhos/cm. TDS ranged from 265 to 1370 with a mean of 756.5 mg/L. The hardness of water samples ranged from 130 to 730 with a mean of 327.22 mg/L. TA, SO_4^{2-} and fluorides ranged from 110 to 620 mg/L, 6.0 to 85 mg/L and 0.06 to 0.68 mg/L.

Fotedar *et al.* (2008) studied the physicochemical parameters in and around Shivkhori Area, Rajouri district, J&K and found all the elements except Si and Al within permissible limits in all the samples according to Bureau of Indian Standard and WHO. Nitrates, sulphates, chlorides, bicarbonates, TDS and total hardness (TH) were also found within permissible limits and hence with respect to all these parameters, the water of the study areas were found safe to be used for human consumption and for agricultural use.

Kaushik *et al.* (2002) assessed the groundwater quality for drinking purpose in the districts Hisar and Panipat of Haryana, on the parameters like pH, EC, Turbidity, TDS, alkalinity, total hardness, calcium, magnesium, sodium, potassium, chloride, nitrate, phosphate, sulphate and fluoride with respect to different land use area viz. residential, industrial, commercial and agricultural and reported that at Panipat groundwater in all the land use zones was fit for consumption, whereas at Hisar, water in agricultural areas was good in quality, but in other areas varied in magnitude of pollution.

Gupta et al. (2009) assessed water quality in seventeen villages of Ambala district and found fluoride concentration higher than the maximum permissible limit of 1.0ppm as per WHO Standards, and higher than the maximum permissible limit of 1.5ppm as per ICMR and BIS Standards.

Garg *et al.* (1998a) conducted an extensive study regarding fluoride distribution in well waters in Jind district and showed that 89% of the studied well water samples had fluoride content more than the permissible limits.

Garg *et al.* (1998b), studied fluoride content carried out in Uklana town, district Hisar. They reported that fluoride content in Uklana town varies from 0.18 to 8 mg/L and fluoride has been found to be negatively correlated with total hardness and positively correlated with total alkalinity.

Singh *et al.* (2006) conducted a study regarding fluoride distribution in groundwater in Pataudi block of District Gurgaon. The mean fluoride concentration in drinking water samples of Pataudi, Haily Mandi and Harsaru villages was 1.68, 3.22 and 1.78 mg/L, respectively. It was concluded from the result that people in the study area were chronically exposed to higher levels of fluoride from drinking water. The latest estimates suggest that around 200 million people, from among 25 nations the world over, are under the dreadful fate of fluorosis. India and China, the two most populous countries of the world, are the worst affected (Ayoob and Gupta, 2006). Groundwater fluoride in high level has been reported (Agrawal et al., 1997; Maithani et al., 1998; and Datta et al., 1999) in all the 31 districts of Rajasthan with a serious health related issue in 23 districts. Muralidharan et al. (2011) found that fluoride was getting enriched at deeper levels. Leachable fluoride concentration was found to be less than 2 ppm at the surface to more than 10 ppm at 140–160 cm depth. This explains the process of fluoride migration through percolating waterfront in each cycle of tank-filling.

There are some reports which indicate that heavy metals pollution including Ni in soils arises as a result of various anthropogenic activities such as continuous use of sewage water (Brar et al., 2002; Kumari et al., 2006; Krishnan et al., 2007), sewage sludge (Singh and Sakal, 2001) and fertilizers (Tiller, 1992; Indra and Sivaji, 2006).

Ahmad and Alam (2003) evaluated the impact of different types of chemical, electroplating, textile and dyeing industry waste water on the river and groundwater. Water samples from the localities located on the side of Yamuna River and other areas in Delhi and industrial effluents of different types of industries were collected and analyzed. Water quality parameters were very poor, except the samples collected from upstream.

Physicochemical and microbiological studies of sugar mill effluent polluted groundwater in Eraiyur area of Permbalur District, Tamil Nadu by

Amathussalam et al. (2002) indicated that EC, TDS, total hardness, BOD, COD and ions level values are on the higher side of permissible limits of WHO standards. Microbiological studies revealed the presence of specific fungal species which are capable of growing in higher concentrations of bicarbonate and nitrates.

Dixit et al. (2003) evaluated the final water supply of four treatment plants and 80 tube-wells at Delhi for heavy metals. The levels of manganese, copper, selenium and cadmium were found marginally above the Indian Standards (IS) specification regulated for drinking water.

Guru Prasad (2003) calculated the Water Quality Index (WQI) of groundwater for Tadepalli mandal of Guntur district, AP and assessed the impact of pollutants due to agriculture and human activities on its quality and found that water of the study area was not safe for human use.

Islam and Gyananath (2002) studied the implications of chemical fertilizers (mean sulphate, phosphate and nitrate concentration) on ground water quality of Nanded. The mean recorded values of sulphate, phosphate and nitrate levels were found 10.26-34.83 mg/L, 0.052-0.194 mg/L and 3.43-11.37 mg/L, respectively. Sulphate and nitrate levels were within permissible limits but phosphate levels higher than the permissible limits.

Gulab Sagar, a sewage polluted pond at the mid of Jodhpur city was studied by Jakher and Rawat (2003) for two parameters - nitrate and most probable number (MPN). The relationship between both the parameters was noted as highly significant. The correlation co-efficient for nitrate and MPN was found to be 0.91 and the empirical parameters were determined to be a = 46.25 and b = 12.48.

Jena et al. (2003) conducted a field survey to study the coastal water quality of the Sagar Island, which plays a decisive role in coastal resource management. Some physicochemical parameters and nutrients of the coastal water during the post-monsoon season were studied. Coastal waters associated with mangroves represented salinity range of 4-7% more than average values of dissolved oxygen (5.84 mg/L) observed in the mangrove patches.

Kaushik et al. (2003) studied the heavy metal pollution in the water of major canals originating from the river Yamuna in Haryana and found all heavy metals except Zn in the Western Yamuna Canal exceeding the maximum permissible limits. Concentrations of the metals were, however, relatively less in the highly eutrophicated waters of Agra canal and Gurgaon canal as compared to that in WYC but Fe concentration were much higher.

Khedkar and Dixit (2003) analyzed the physicochemical characteristics of the wastewater generated by the vast population of Amravati and found the majority of the parameters within permissible limits, but the sodium concentration in the wastewater exceeded the standards recommended by CPCB or WHO.

Manjapa et al. (2003) analyzed 61 different bore-well samples, selected from different areas of Davanagere taluk and found 26% of the samples contain fluorides less than safe limit prescribed by BIS and 11.5% of the samples are found to contain more than safe limit. Further, 16% of the bore-well samples analyzed were found to contain more than safe limit prescribed by BIS. The values of fluorides and nitrates observed in different samples were in the range of 0.19-2.06ppm and 0.08-308ppm, respectively.

Matkar and Gangotri (2002) studied the sugar industrial effluents toxicity to aquatic fauna and human health and found pH of the effluent at 4.00, BOD 43000 mg/L and COD 89760mg/L which is beyond the tolerance limit of the water, causing shifting of the algal forms towards more tolerant zone, leading to decrease in biodiversity. Total dissolved solids and total suspended solids were also considerably high.

Maya (2003) studied the bacterial quality of water along with seasonal analysis of certain important physicochemical parameters of some temple tanks in Kerala. The overall analysis indicates poor quality of water of temple tanks with organic pollution and faecal contamination.

Meenakumari and Hosmani (2003) assessed the total coliform and *E. coli* level of ground water (open-wells, bore-wells) in various parts of Mysore city and found the values of total coliform from 3 to 2400/100 ml. The high values of *E. coli* were observed in north and east parts of city. The study concluded that large amount of unplanned release of sewage water into subsurface water is largely responsible for bacteriological pollution of ground water in area.

Patel et al. (2003) carried out a survey around major industrial cities to study the level of contamination in water-soil-plant system and found all major polluting elements in the industrial effluents. Further, TSS, COD, BOD values were found above the standard permissible limits for irrigation. The well-water from Bharuch site was found contaminated with Cr and Mn whereas Ankleshwar site contained Fe above the standard limit for irrigation, besides salinity and alkalinity hazards.

Hydro-geochemical investigations carried out by Sujatha and Rajeswara (2003) in the south-eastern part of the Ranga Reddy district, Hyderabad to assess the quality of groundwater for domestic and irrigation purposes showed the concentrations of NO_3^-, Cl^-, and F^- ions above the permissible limits and found that the extensive use of fertilizers and large-scale discharge of municipal wastes into the open drainage system of the area, may be the possible cause.

Pathade et al. (2003) Studied microbiological analysis of drinking water samples from some hotels and schools in Karad, Maharashtra for coliforms and water borne enteropathogenic bacteria. Enteropathogenic bacteria like *Pseudomonas aeruginosa, Klebsiella pneumoniae, Staphylococcus aureus, Shigella* species and *E.coli* were commonly found in school and hotel drinking water samples. More than 40% samples showed more than 240 coliforms/100 mL and pathogenic isolates showed resistance to many antibiotics of common use.

Purandara et al. (2003) studied samples of water from Bellary nala (Belgaum city) which was once a freshwater stream and now turned into a waste stream and revealed that the surface water is completely deteriorated as indicated through dissolved oxygen concentration. It was also observed that in industrial patches and adjoining areas of nala, increase in salinity contents may turn into saline water in years to come if not properly cared.

Rawat (2003) studied the direct and indirect factors affecting microbial fauna of the two water bodies of Jodhpur region and found that the water to be unsatisfactory for drinking and other purposes throughout the year. The coliform number was found maximum in June and July at Gulab Sagar and Takhat Sagar respectively, with a trend of the fall of coliform number in winters, rise in summers and again maximum in rains.

Sharma et al. (2003) studied the water quality of Hathli stream in Hamirpur district of Himachal Pradesh in lower Himalayan region and revealed that the water in the stream is heavily polluted as BOD, TDS, hardness and alkalinity exceed the permissible limits and also observed presence of coliforms in excessive numbers.

Sharma and Verma (2003) analyzed water samples collected from natural springs in Hamirpur area of Himachal Pradesh and found the physicochemical parameters within the maximum permissible limits of drinking water standards. However, low fluoride and iron is observed in all the spring water samples. The study also revealed that water of the area is very hard and highly alkaline and is dominated by bicarbonates, calcium and magnesium.

Sikdar and Banerjee (2003) found arsenic above permissible limit of 0.05mg/L in groundwater samples in parts of seventy-three blocks and eleven municipalities of eight districts of West Bengal and discussed that the hypothesis of geological source of arsenic has certain drawbacks and also highlighted the alternative anthropogenic sources of arsenic.

Srinivas et al. (2002) studied the various physicochemical characteristics of water samples collected from in and around dumping yards in the Visakhapatnam city found significant increase in the iron concentration.

A study by Sujatha (2003) found the concentrations of fluoride in the groundwater vary from 0.7 to 4.80 mg/L and from 0.4 to 4.20 mg/L during the pre and post-monsoon seasons respectively, in the Ranga Reddy district, Andhra Pradesh. By contrast, the fluoride concentration in many places was relatively high during the post-monsoon period. This indicates contamination of groundwater from surface pollutants.

Kumar and Sinha (2010) studied the twelve drinking water quality parameters of the twelve water samples of hand pumps at Moradabad and found the water quality of study area polluted.

Tripathy (2003) analyzed the groundwater samples in and around Bhanja Bihar, Orissa to determine total dissolved solids and concentration of major ions. Analysis results found the groundwater fit for human consumption and the Cl^-/HCO_3^-, and Mg^{++} / Ca^{++} values indicated that the aquifers are free from any salt water ingress from the sea as is the case with several localities along the coast.

Umar and Absar (2003) collected twenty-nine dug-well samples from the Gambhir River basin in the Bharatpur District of Rajasthan for hydro-geochemical study to understand the sources of dissolved ions and assess the chemical quality of the water and found that the groundwater has chemical composition within the permissible limits suggested for drinking water. Nitrate was found higher than the acceptable limit in some samples, due to the use of fertilizers.

The chemical characteristics of groundwater in Malaprabha Sub-basin of Belgaum District, Karnataka have been studied by Varadarajan and Purandara (2003) during the pre-monsoon and post-monsoon seasons to evaluate the suitability of water for domestic and irrigation purpose and found the quality of groundwater in the upstream region of the sub basin quite acceptable for both the uses, whereas in the downstream region various parameters exceeds the acceptable limits due to excessive irrigation by excess application of fertilizers and pesticides. In addition to this fluoride was observed in excess along the downstream region of the sub basin.

Physicochemical studies by Yadav et al. (2003) regarding the water quality assessment of some villages of Behror tehsil indicated the high values of inorganic salts, nitrate, fluoride and hardness showing the water quality totally unfit for drinking purpose and harmful for the health of the consumer.

Sabal et al. (2008) analyzed the 100 groundwater samples collected from 25 villages of Amber tehsil of Jaipur district for physicochemical parameters like Fluoride, pH, electrical conductivity, total dissolved solid, total hardness, calcium, chloride and alkalinity and revealed considerable variations in the chemical composition of water samples and Fluoride concentration varies from 0.91 to 4.20 mg/L.

Analysis of groundwater samples collected from various locations of Bhavnagar region by Mishra et al. (2009) revealed that turbidity, manganese, zinc and copper were within permissible limits but TDS, total hardness, chloride, fluoride and chromium were observed beyond permissible limits in some samples and iron in almost all the samples and concluded that groundwater in Bhavnagar region requires precautionary measures before drinking to avoid adverse health effects.

The investigations by Dayal (1992) indicated a high degree of pollution in groundwater of Agra city. Though much of the variables were within the standard limit of potable water, a few heavy metals recorded a concentration much beyond the permissible limits set by the WHO (1984). The water sources around septic tanks and sewage channels showed a high contamination of coliforms.

Twenty two ground water samples from Hubli city, Dharwad district Karnataka analyzed by Hegde and Puranik (1992) for Fe, Mn, Cu, Ni, Pb and Cd indicated high concentrations probably because of industries and highway contamination.

Study done by Kaur et al. (1992) to investigate seasonal variations of various water quality parameters and to investigate correlations amongst Bicarbonates, chlorides, electrical conductivity, pH, sulphates and total hardness show a downward trend in their concentrations from April to June (summer season) whereas their concentrations increase during August to October due to monsoon rains, indicating potential for contamination due to acid rain.

Study by Anand et al. (2006) revealed the impact of diverse anthropogenic activities as well as the monsoon effects on the bacterial population of river Yamuna in Delhi stretch. Microbial population contributed mainly through human activities prevailed there.

The degree of trace elements pollution and the suitability of groundwater for drinking purpose were assessed by Barik et al. (2006). The concentration of lead and iron was found above maximum permissible limit.

Experiments carried out at by Das et al. (2006) indicated that electrical conductivity has a linear relationship with Total Dissolved Solids, which is validated by the findings at various other lakes throughout the world.

High fluoride contamination and high concentrations of SO_4^{2-} has been observed by Dutta et al. (2006) in the groundwater of areas having ancient alluvial red soil and Precambrian metamorphic rock complex basement in the Kapili-Jamuna sub-basin.

TDS and total hardness was found above the desirable limit by Jadeja et al. (2006) in all the groundwater samples collected from Dharampur industrial area Porbandar city.

From the analytical data of physicochemical parameters, Mishra et al. (2005) found river and groundwater contaminated and indicated that it might be due to the industrial and municipal effluents.

Pawar et al. (2006) analyzed the bore-well and dug-well water samples from a highly polluted industrial area of Nacharam, Hyderabad for physicochemical parameters and found those unfit compared with drinking water standards.

Natural spring-water and dug-well water from Lote Industrial Area and nearby villages was analyzed by Raje et al. (2005) and found the groundwater with varying degrees of trace metals contamination that may cause a serious health problem to domestic animals and human beings.

Ground water quality of industrial area of Kishangarh, Rajasthan was studied by Sharma et al. (2005) for various physicochemical parameters seasonally and from the study it was clear that these parameters increase with the addition of marble slurry leading to deterioration of the overall quality of the groundwater.

Total coliform and faecal coliform bacteria in water from two channels at Okhla in the southern part of Delhi monitored by Mohapatra et al. (1992) were found to be infected with high count of coliform bacteria. Total coliform and faecal coliform counts had the highest values in the month of March while the lowest values were obtained in December.

Patel (1992) found groundwater quality of the Garhkalika, Bharathari caves, southern side of Runmuktheswar and Solah sagar region, Nagar Kot Ke Rani region, Rudhra sagar and Jaisinghpura region and eastern side of Hira and Binod Mills regions unsuitable for irrigational purpose as groundwater samples of the these region indicated high salinity.

Water samples from 30 rivers in northern and northeastern hilly states of India analyzed by Pathak et al. (1992) for bacteriological and physicochemical parameters along with metals and pesticide residues indicated 34% of samples with > 50 coliforms/ 100 ml, 24% of samples demonstrated > 50 thermo-tolerant (faecal) coliforms/100 ml. Among the metals, iron was found to be above maximum permissible limits in the rivers of all the states, while manganese was found to be above the maximum permissible limits in the rivers of Tripura and some northern states.

Sharma et al. (1992) analyzed Fifty three water samples from different wells used for irrigation within Chambal command area and found that the majority of waters are high in salinity and sodium.

Dissolved nutrients estimated by Shibu et al. (1990) in the surface and bottom waters of five selected stations of the Evaravur Lake during February 1987 to January 1988 revealed distinct seasonal variations. Rainfall and land drainage play significant roles in the nutrient economy, particularly NO_3-N and SiO_4-Si of this water body. Abnormally high values of PO_4-P indicated extremely polluted condition at the retting zone of the lake during the pre-monsoon season.

Water of River Ganga was studied by Shukla et al. (1992) at four sampling sites at Ghazipur, U.P., from May 1987 to April 1988 and found high bacteriological count, besides depletion in the dissolved oxygen, and increase in ECE, BOD, COD, pH, nitrate N, phosphate P, sodium, potassium and calcium contents.

Physicochemical and bacterial parameters of 23 bore wells and dug wells of 23 villages of Challapalli Mandal were assessed by Somasekhara et al. (1992) and found that there is high incidence of fluoride.

Hydrological and sediment transport characteristics are two main fluvial parameters affecting the Aggradation/degradation behavior of the river systems (Sinha, 2005). Mass bathing in sacred water bodies is an age-old ritual in India. Study by Semwal and Akolkar (2006) deals with water quality assessment of rivers in Uttaranchal, in view of their religious importance and ecological sustainability. The physicochemical water quality in most of the rivers of Uttaranchal remained unchanged except total dissolved solids ranged from

90.23 to 121.33mg/L, total suspended solids varying from 126.5 to 236.5mg/L and total alkalinity of 37.0 to 96.0mg/L. Significant levels of sulphates (1.66 to 20.0mg/L) were also observed at the religious places. Traces of iron, zinc and copper metals, Endosulfan, Dieldrin and DDT in water and sediments have been observed in clean water quality stretches. Krishnaswami and Singh (2005) observed total dissolved solids in Himalayan Rivers ranged from 35 to 151 mg/L.

Singh *et al.* (1962) reported that the average fluoride content of 60 water samples of village Bajekhana of district Bathinda ranged from 2.4 to 16.2ppm with a mean of 8.4ppm. From an analysis of large number of well-waters of Bathinda district, they reported a mean value of 0.30 to 14.00ppm of fluorine in these waters.

A study conducted by Joia et al. (1978) on residues of DDT and HCH/BHC in Wheat Flour in Punjab detected 124 and 116 samples affected, respectively out of 140 samples of wheat flour.

Goyal et al. (1981) indicated that out of total groundwater reservoir in Punjab, 76.5% groundwater is fit and the rest showing varying degree of salinity. The fit water zone being the main attraction for the intensive industrialization and urbanization in the state is polluted by unsatisfactory disposal of industrial waste and lack of proper sewerage facilities. High concentration of nitrates, cyanides, and trace elements has been reported in the groundwater samples from Ludhiana town.

Brar *et al. (*1984) have studied hydro-chemistry of ground water of *Bhawanigarh block* in Sangrur district of Punjab and reported fluoride concentration in the study area in the range from 0.28 to 4.00 mg/L except one sample from the *Bimberi village* exceeding the critical limit of 1.0 mg/L and concluded that there is no problem of fluoride hazard in the ground water of the *block.*

A study was undertaken by Singh et al. (1990) on the effects of sewage irrigation from Ganda Nallah (Hakimwala drain) and Tung Dhab Drain on the soil and crops in Amritsar during 1986 to1990. It was observed that copper and zinc accumulated in the soil with prolonged sewage irrigation. The contents of these two heavy metals were found up to 270μg/g and 412.8μg/g, respectively, which were very high for cultivated lands.

Uranium estimation has been carried out by Singh et al. (1995) for some water samples collected from Bathinda and Amritsar showed the range of Uranium concentration present in water samples taken from Amritsar was 17.87

± 0.18 to 20.23 ± 0.20 ppb while that in Bathinda samples was 11.71 ± 0.15 to 113.70 ± 0.46ppb.

Dhillon et al. (1997) found four-fold increase of selenium content in the areas of Hoshiarpur and Jalandhar districts of Punjab due to following of rice-wheat cropping pattern as compared to when maize-wheat cropping pattern was practiced.

Sood *et al.* (1998) reported that electrical conductivity of water samples collected from 88 villages of Talwandi Sabo tehsil of Bathinda district, varied from 0.55 to 13.74 dS/m. Soluble carbonates, bicarbonates and residual sodium carbonates of these waters varied from 0.0 to 3.2, 2.0 to 17.7 and 0.0 to 14.6 me/L respectively.

Results of a study conducted by Khurana *et al.* (2003), in districts Ludhiana, Jalandhar, Amritsar and Sangrur showed elevated levels of DTPA (Diethylene triamine penta-acetic acid) extractable heavy metals in sewage irrigated soils as compared to tube-well irrigated soils.

Dhillon et al. (2004) has reported selenium in groundwater of north-east parts of Punjab in the range of 2.54 to 69.53 µg/L (safe limit is 10 µg/L), especially in parts of Hoshiarpur and Nawanshahr districts.

A study by Kumar (2005) indicated higher pesticide contents as well as higher concentrations of mercury, chromium and selenium in groundwater of Chamkaur Sahib of Rup Nagar district as compared to tap water while Battu (2005) found residents of Talwandi Sabo, Bathinda using canal water brought from the other areas instead of using groundwater and assessed that the pesticides in all the samples were below detection limits. However, Mathur et al. (2005) found high level of pesticide residues in human blood samples taken from villages of Punjab.

According to a study conducted by Sharma et al. (2005) in Jalandhar and Moga districts to find out the awareness level of the farmers regarding the implications caused due to excessive use of pesticides, indicated that 28% of the respondents were not aware about instructions written on pesticide containers, 64.5% respondents were not aware about recommended dose of inputs, 48.5% respondents were unaware of the need to keep pesticides in original containers and 54% were careless about their safe storage. Majority of the farmers (75.5%) did not dispose off empty containers; rather they were reusing them in household activities. Persistence of some pesticides was known to a majority of farmers (67%), but only 36% respondents were aware about the hazardous effects of their excessive use. About 54% of farmers were unaware of the ill effects like respiratory and skin diseases and allergies caused by pesticides in human beings.

Minakshi et al. (2006) studied the groundwater quality of Rupnagar district of Punjab and found that Kharar block has 46.7 percent of the total geographic area under marginally sodic water as compared to the Anandpur Sahib block with 8.5 percent of its area under marginally sodic waters.

A study was undertaken by Saini et al. (2006) to characterize underground irrigation water collected from 84 different tubewells of south-western zone of Punjab, found to have 7.19, 7.16, 6.99 and 1.48 mg/L of Ca+Mg, Cl⁻, HCO_3^- and CO_3^{--}, respectively. On the basis of EC, 64.3% were rated as fit, 28.6 marginal and 7.1% unfit for irrigation. On the basis of RSC, 5.12% samples fell in the category of fit and rest 28.6 and 20.2% were marginal and unfit, respectively, for irrigation.

Hundal et al. (2007) assessed the arsenic contents groundwater samples of north-eastern, central and south-western districts of Punjab and found 41%, 46% and 54% samples with high arsenic (>50ppb) contents, respectively.

Tiwana et al. (2007) indicated that residue of chemical pesticides in human beings, milk, water, vegetables and other food products are at levels which are dangerous for human health. After the ecstasy of the green revolution, Punjab is now to battle with residual effects of extensively used chemical pesticides in air, water, environment and food products.

Industrial and agricultural activities in urban areas of Punjab (Saxena et al., 2007) have led to a considerable increase in heavy metal levels in different environmental compartments, especially in soils over the course of recent decade.

A qualitative water quality survey was carried out by Verma et al. (2007) to diagnose the salinity or sodicity hazards in ground waters in relation to their suitability for irrigation in Faridkot district (Punjab) having two blocks namely Faridkot and Kotakpura. The total concentration of soluble salts in ground waters varied from 0.31-7.53 dS/m. On splitting the analytical data block-wise, it was observed that groundwater of Faridkot block pose high salinity hazards than that of Kotakpura block.

A study by Thakur et al. (2008) found level of As, Se, Hg in ground water at Talwandi Sabo more than the permissible level where as at Chamkaur Sahib the level of Se were above the permissible level. Similarly, As, Se were also above permissible level in tap water at Talwandi Sabo whereas as in

Chamkaur Sahib only As levels were more than the permissible limit in tap water.

Khurana and Bansal (2008) evaluated the effects of irrigation with contaminated sewage water in soils and its accumulation in crops in four major industrial towns of Punjab namely Jalandhar, Ludhiana, Amritsar and Mandi Gobindgarh.

Study of drinking water qualities in the areas of Talwandi Sabo and Chamkaur Sahib by Thakur et al. (2008) revealed higher levels of heavy metals such as As, Cd, Cr, Se, Hg and pesticides such as heptachlor, ethion, and chloropyrifos were also higher in samples of drinking water, vegetables, and blood in Talwandi Sabo as compared to Chamkaur Sahib.

Mittal and Sharma (2008) conducted a study to assess the drinking water quality at Moga, Punjab as this area consumes highest quantity of fertilizers and pesticides. It was found that the major physicochemical parameters were within the permissible limits but certain parameters like total dissolved solids, electrical conductivity and magnesium contents were above WHO permissible limits at almost all the places in Moga city. Almost 60% of water samples were found polluted.

Samson et al. (2010) studied the classification and water quality assessment parameters like dissolved oxygen, conductivity, pH, turbidity, total and suspended solids (SS), chemical oxygen demand, and Secchi disk transparency (SDT) of Harike wetland (Ramsar site) in India, converging on two rivers, Beas and Sutlej, using satellite images and correlations were established between turbidity and SS, SS and SDT, and total solids and turbidity.

Mercury was found higher than permissible limits by Thakur et al. (2010) in 84.4% samples in 35 villages of three districts of Punjab. Heptachlor, chlorpyriphos, β-endosulfan, dimethoate, and aldrin were found to be more than permissible limits in 23.9%, 21.7%, 19.6%, 6.5%, and 6.5% in groundwater samples respectively. A larger proportion of children in target area were reported to have delayed milestones, language delay, blue line in the gums, mottling of teeth and gastrointestinal morbadities.

Chapter-3

PATIALA - THE DISTRICT UNDER STUDY: AN OVERVIEW

Patiala district is one of the famous princely states of erstwhile Punjab. Famous for 'peg', 'pagri', 'paranda' (tasselled tag for braiding hair) and 'Jutti' (footwear), joyous buoyance, royal demeanor, sensuous and graceful feminine gait and aristocracy, Patiala presents a beautiful bouquet of life-style even to a casual visitor to the city. Forming the south-eastern part of the state, it lies between 29°49' and 30°47' north latitude, 75°58' and 76°54' east longitude. Patiala is an important district in the Malwa zone of the State. It is surrounded by Fatehgarh Sahib and Rupnagar districts and the Union Territory of Chandigarh to the north, Fatehgarh Sahib and Sangrur districts to the west, Ambala and Kurukshetra districts of neighboring Haryana state to the east and Kaithal District of Haryana to the south west. Patiala, the headquarters of the district administration is linked to Chandigarh (70 km) and New Delhi (253 Km). Patiala district accounts for 6.7 percent of the total population of the state in about seven percent of its area. With a total population of 16.34 lacs (Census, 2001) the district ranked 6th in the state after Amritsar, Ludhiana, Gurdaspur, Sangrur and Jalandhar. With regard to distribution of the rural population in 2001, around 40percent in the district live in villages with total population 1000-1999, 30percent in villages with the total population 2000-4999, 22percent in villages with total population 500-999, six percent in villages with total population below 499 and rest in villages with population more than 5000. Classification of working population in Patiala district shows that 22percent are cultivators, 18percent are agricultural laborers, three percent are engaged in the household industry and a whopping 57percent are other type of workers.

The Patiala district is divided into five sub-divisions (tehsils) namely Patiala, Nabha, Ghanaur, Rajpura and Samana comprising eight-community development blocks viz. Patiala, Nabha, Sanour, Bhunerheri, Rajpura, Ghanaur, Samana and Patran for the purpose of administration. Patiala is the largest tehsil while Nabha is the smallest. Patiala district has 11 towns and 988 inhabited villages (Bhunerheri, 160; Ghanaur, 126; Nabha, 171; Patiala, 114; Patran, 75; Rajpura, 154; Samana, 81 and Sanour, 107).

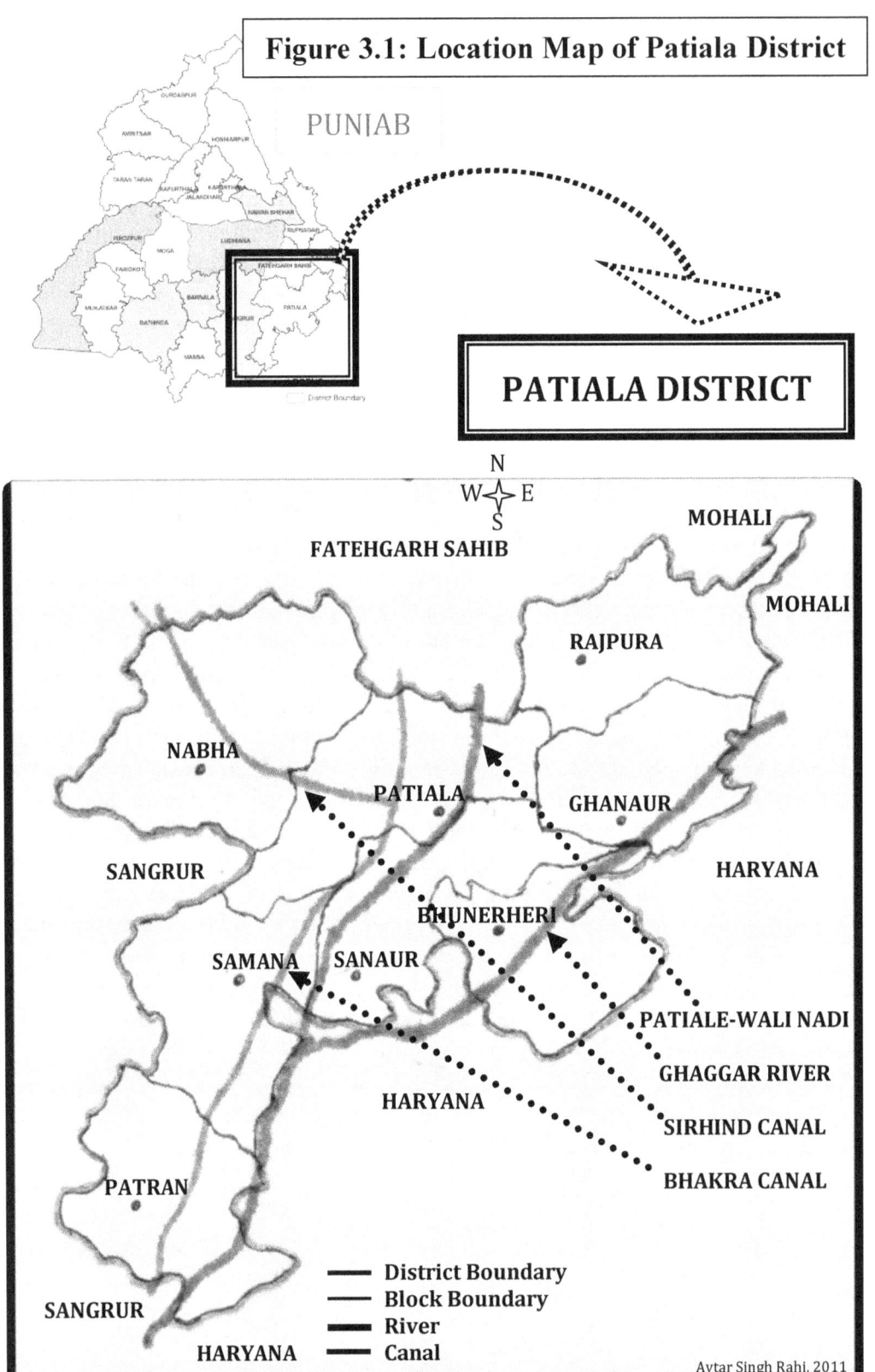

Figure 3.1: Location Map of Patiala District

PUNJAB

PATIALA DISTRICT

N
W E
S

FATEHGARH SAHIB

MOHALI

MOHALI

RAJPURA

NABHA

PATIALA

GHANAUR

SANGRUR

HARYANA

BHUNERHERI

SAMANA   SANAUR

PATIALE-WALI NADI

GHAGGAR RIVER

HARYANA

SIRHIND CANAL

BHAKRA CANAL

PATRAN

District Boundary
Block Boundary
River
Canal

SANGRUR

HARYANA

Avtar Singh Rahi, 2011

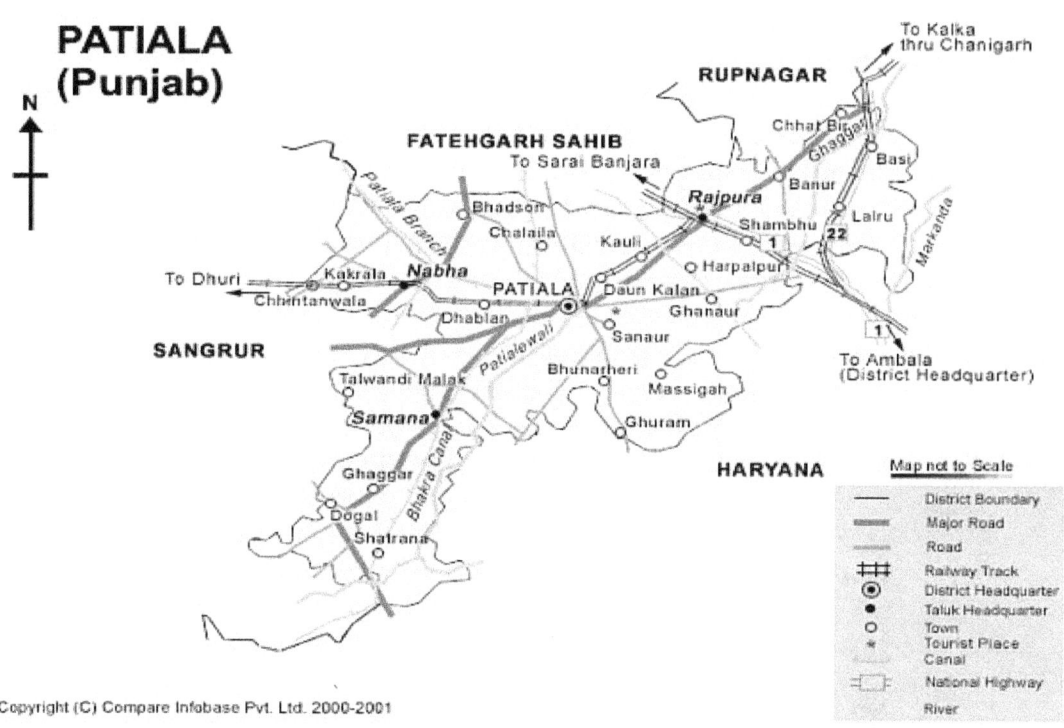

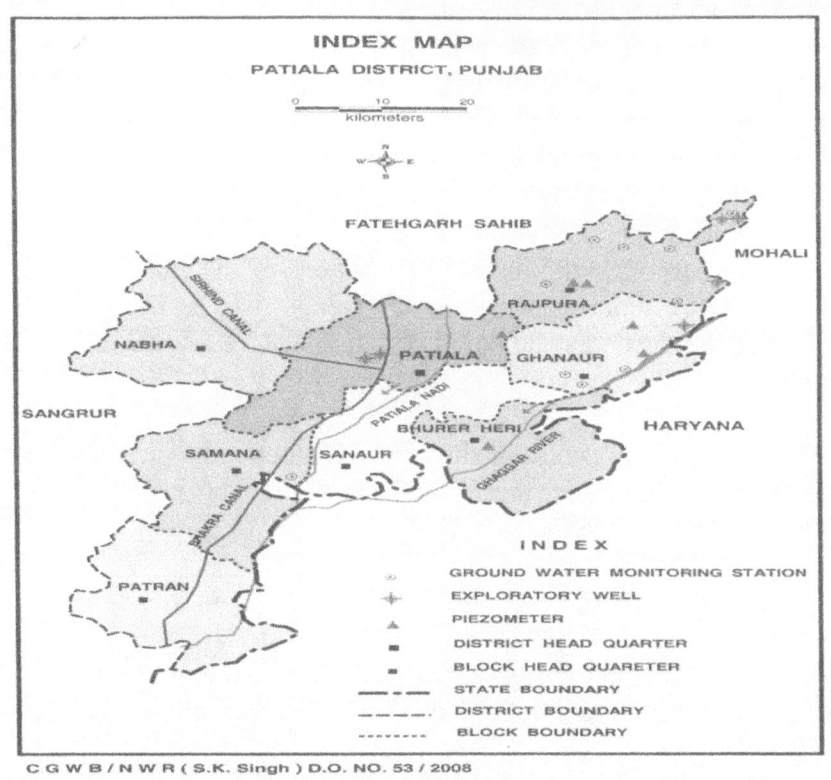

Patiala has been cultural and academic center of northern India. Educationally, Patiala is in the forefront. Patiala was the first town in this part of the country to have Degree Collage-the Mohindra College-in 1870. This city has the credit of setting up the first printing press in Punjab; the famous Printing Press of Munshi Naval Kishore and for manufacturing the first Punjabi Typewriter. Patiala is an important seat of learning with Punjabi University, Rajiv Gandhi Law University, Thapar University, National Institute of Sports, Rajindra Hospital, Government Medical College, Dental College, Ayurvedic College, Engineering College and many others. Punjab Pollution Control Board, Punjab Public Service Commission, etc. are also the part of Patiala.

It is predominantly a rural district, where an overwhelming 65% lived in rural areas. Agriculture is the single most important economic activity in the district. 81% of geographical area out of 3290 square kilometers in Patiala district is cultivable. 93% of the area is irrigated through tube-wells and 3% by canals.

3.1   TOPOGRAPHY

The undulations or smoothness of land surface along with its slopes and surface materials exert a strong influence on human activities of an area and Patiala District is no exception. The district forms a part of the Indo-Gangetic plain. It is composed of materials deposited by rivers during recent geological past. These deposits belong to the quaternary era and thus are either equal to or less than 1.8 million years in age. The district is a level plain with gentle slopes. It has highest elevation of 320 meters near village Nimbnia located in extreme east-central part of Rajpura tehsil. The lowest elevation of the district has been observed near village Kutbanpur located in southern part of Samana tehsil. Its land surface slopes in the north-east-south-east direction with a gentle gradient of about 0.8 meter per kilometer. Direction of flow and sinuous courses of streams flowing through the district are evident of the same. The district has mean elevation of about 265 meters. It extends over a distance of about 91 kilometers in the east-west direction and nearly 113 kilometers in the north-south direction.

The district area is occupied by Indo-Gangetic alluvial plain and consists of three types of region viz. the Upland plain, the Cho-infested Foothill Plain and the Floodplain of the Ghaggar River.

The Upland Plain: This terrain unit singly covers about 80 per cent of the total area of the district. Leaving aside a small cho-infested tract in the north-east and a narrow belt running along river Ghaggar in the east, south-east and south, the whole of the district is covered by this unit. The smoothness of its surface is distributed by the presence of sand dunes in its western part. It may be further divided into two sub regions; Eastern Upland Plain and Western Upland Plain.

The **Eastern Upland Plain** covers the most of Rajpura Tehsil and eastern as well as southern areas of Patiala Tehsil. This terrain unit is flat and featureless. Being alluvial in origin, its soils are well drained and fairly fertile. Sub-soil water conditions over here are favorable for the development of tube-well irrigation. Its flat land surface coupled with easily workable loamy soils and favorable groundwater situation provide excellent conditions for agriculture. Almost the whole of its cultivable land is subjected to agriculture.

The **Western Upland Plain** occupies the whole of Nabha and Samana tehsils and north-western and western parts of Patiala Tehsil. This part of the upland plain is superimposed by sand dunes at various places. The energetic and hardworking farmers have installed tube-wells over these sand features and have leveled them with their tractors for practicing irrigation. Some of them, however, still are visible in the form of sandy soils. This terrain unit is also cultivated to its near limits. However, its sandy parts are less productive.

The Cho-infested Foothill Plain: This terrain unit occupies nearly 4 per cent of the area of Patiala District. It covers eastern most part of Rajpura Tehsil. Its western boundary is demarcated roughly by the Chandigarh-Ambala Highway, whereas its southern boundary is marked by a line joining Lalru town with Rani Majra village. Its eastern boundary coincides with the district boundary. Its elevation ranges between 290 and 320 meters. Thus, it is located at a higher elevation than the rest of the district with a steeper gradient. It slopes at the rate of 2.5 meters per kilometer (as against the district's average of 0.8) in the north-east-south-west direction. A number of seasonal streams, locally known as *chos*, traverse through this unit, which is its special feature. They originate in the Shiwalik Hills and after traversing this region join either the Ghaggar River or any of its tributaries. These hill torrents bring floods during rainy season and spread sand on the neighboring lands. The soils of this region are lighter, porous, and hence are less fertile. A notable part of its land is under the beds of seasonal streams and thus is not available for cultivation. Underground water over here is deep and inadequate. Agriculturally, this is the least developed part of the district.

The Floodplain of the Ghaggar River: This terrain unit covers nearly 16 per cent of the total area of the district. It runs in a narrow belt all along the river. Being a mighty river in the past, its floodplain extends up to 6 kilometers on both sides of the river at places. However, its current floodplain is not more than 3 kilometers in width on either side of the river at any place. Its present floodplain is flooded during heavy rainfall. It results in damaging crops in low lying areas. However, in relatively higher areas, its floods bring boom for rice cultivators. Floods deposit silt and enrich its soils. Its soils are heavier (silty-loams) and area ideal for the cultivation of rice. Sub-soil water in this tract is relatively close to the surface and is found in adequate amount. This fact has encouraged the development of tube-well irrigation.

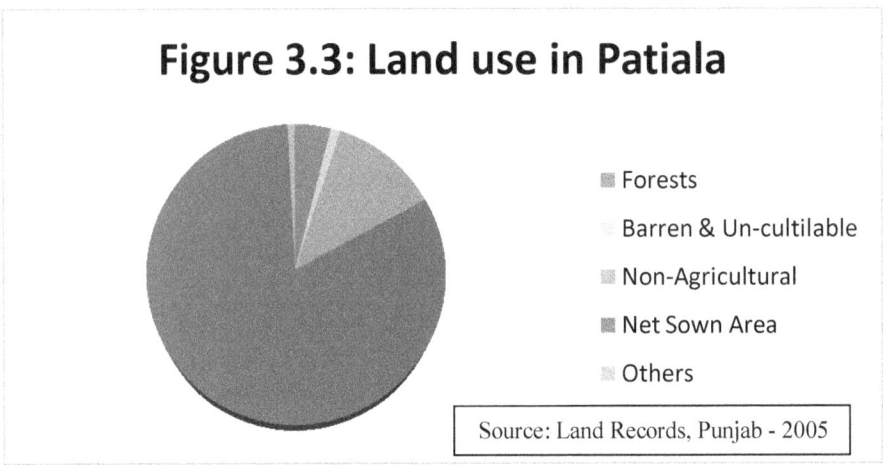

Figure 3.3: Land use in Patiala

- Forests
- Barren & Un-cultilable
- Non-Agricultural
- Net Sown Area
- Others

Source: Land Records, Punjab - 2005

3.2   SOILS

The soils in the sand dune area are sandy to loamy sand in texture with low available moisture capacity 2.3 to 5.0 cm/100 cm profile. Soils in the inter-sunal plains are relatively heavier (sandy loam to sandy clay loam) with moderate available moisture capacity 4.8 to 18.0 cm/100 cm. The soils in the levelled plains are characterized by loam to clay loam texture occasionally heavier profiles with clay texture are observed locally. Low lying areas in these terraces are salt affected. The soils in the old channels of Patialewali Nadi and Choa Nadi are very fine textured (clay and silty clay) and they contain lot of shrinking and swelling type of clays. These areas also experienced seasonal flooding resulting in high water table and are mostly salt affected. The soils of the Ghaggar floodplain are variable in texture. The soils of the river sand bars are sandy and those of the surrounding floodplains are generally sandy loam to silty clay loam in texture. The stratified soils are calcareous in nature. The soils in the undulating plains (near the Shivalik) are having 2-15 per cent slope and are sandy loam to loam in texture. These soils are susceptible to erosion by choas during monsoon season.

3.3   IRRIGATION

Patiala, being predominantly agricultural district, the well-being and prosperity of the tiller depends upon, to a large extent on the irrigation. There is a positive correlation between agricultural production and irrigation facilities. Besides rainfall, surface water (Rivers, Canals) and groundwater (Tube-wells, wells) are the sources of irrigation in the district. A lot of efforts are made to bring more areas under irrigation and a number of irrigation schemes are being taken up.

3.4   RAINFALL

The average annual rainfall in the district is 604.3mm. The rainfall in the district in generally increases from southwest towards the northeast and varies from 508.2mm at Dhanetha to 674.2mm at Patiala. About 70 percent of the annual normal rainfall in the district is received during the south-west monsoon period, July to September, July and August being the rainiest months. Some rainfall is also received during June and October and in the cold season.

3.5   SURFACE WATER RESOURCES

River Ghaggar and its tributaries form a major natural drainage system of the district. Apart from this, some canals too flow through it for considerable length.

Ghaggar River: Ghaggar is the most important water channel of the district. It was once a mighty river with Yamuna and Satluj as its tributaries. At that time, it formed a part of the Indus system. But the uplift of Yamuna-Satluj divide in the past is considered to have shifted Yamuna to the east and Satluj to the west, leaving Ghaggar as a misfit river. Now it loses itself in the sands of the Thar Desert.

River Ghaggar originates in Sirmaur District of Himachal Pradesh flows through the Himalayas for some distance ultimately to enter Punjab plains near Panchkula in Haryana. Flowing further down for a short distance, it enters Patiala District near Mubarakpur village, continues its march in an overall north-east-south-west direction and passes through the eastern, south-eastern and southern parts of the district. It traverses approximately 110 kilometers of distance within Patiala District. After flowing for a distance of about 80 kilometers, it leaves the district near village Jalalpur of Patiala Tehsil and re-enters near village Rattan Heri in Samana Tehsil. Thereafter, it flows towards south along the eastern boundary of tehsil Samana up to Rasauli village from where it turns towards south-west and finally leaves the district near Gulahar village of Samana Tehsil. On the way, it is joined by a number of seasonal streams, such as Dangri Nadi, Markanda River and Patialewali Nadi at various points of its course.

Ghaggar is essentially a seasonal stream. It contains a streak of water in its upper course throughout the year, but remains dry in its lower parts during most of the year. However, during rainy seasons, it is full of water. It is flooded during heavy rainfall years and affects the life and property of the people settled in its floodplain. It has a braided course in its upper parts where its channel is shallow and wide. Its depth ranges from 2 to 4 meters in Rajpura tehsil. But in Patiala and Samana tehsils, the depth of its channel increases from 7 to 9 meters. Along with it becomes narrow.

Dangri Nadi: After originating in the outer Himalayas, it flows for most of its length through Ambala District of Haryana. It passes through Patiala District at two places (i) south-eastern tip of Rajpura tehsil for about 8 kilometers and (ii) south-eastern Patiala tehsil for nearly 30 kilometers. Further down, it merges with Ghaggar River. It is a seasonal steam. Although it traverses only a distance of 30 kilometres in Patiala tehsil, yet is known for severe floods. It flows through a narrow and entrenched channel of 2 to 3 meters deep in Patiala tehsil and wider and less deep in Rajpura tehsil.

Patiale-wali Nadi: This seasonal stream originates from the Shiwalik Hills. After traversing some distance in Kharar tehsil of Rupnagar District, it enters Patiala District near Machhli Kalan village in Rajpura tehsil and passes through northern parts of Rajpura, then enters Patiala tehsil from the north, touches Patiala city and moves south-east to join river Ghaggar near village Ratta Khera. It traverses about 95 kilometers in the district.

Sirhind Choa: Sirhind *choa* is another seasonal stream of the district. It originates near Sirhind town from the rain waters of the area. Jainti Devi-ki-Rao, another seasonal stream, which terminates in this area, might be partly contributing to its origin. It flows through central parts of Nabha tehsil and leave the district near Chhintanwala village after traversing about 80 kilometers in the district. It remains dry except during rainy season.

Jhambowali Choi: This seasonal stream originates near Chanarthal Kalan in district Fatehgarh Sahib by taking rain waters of the area and flows southward through Patiala and Samana tehsils before falling into Bhupinder Sagar lake. Beyond this lake, it leaves Patiala District to join river Ghaggar. It has a total length of about 95 kilometers within the district.

Tangauri Choi: This drainage line originates from the sewerage and rain water of the Chandigarh City and areas around it. After flowing for some distance through Kharar tehsil of Rupnagar District, it enters Patiala District near Kalauli village in Rajpura tehsil. Flowing further down to the south-west, it touches outskirts of Rajpura town after which it runs south to join river Ghaggar near village Sural Kalan. It is seasonal in character and is full of water during rainy season. However, it does not dry completely during other seasons. It contains a streak of water during other seasons too because of the disposal of sewerage waters of Chandigarh City into it. Since its water is charged with sewerage contents, it is highly productive for crops and the local farmers pump it at many places for irrigation.

Besides the above, a few other seasonal and shorter drainage lines also originate within or outside the district, flow for short distance within in and are lost.

Canals: Bhakra Canal runs through the centre of the district in north-south direction for a distance of about 110 kilometers. Sirhind Canal is another drainage line of this type. It runs for a length of about 32 kilometers through south-eastern parts of Nabha tehsil. These main canals, through their branches, distributaries and minors, provide irrigation water to various areas of the district.

Other Water Bodies: A lake and innumerable ponds are the other water bodies of the district. Bhupinder Sagar Lake is located in southern part of Samana tehsil. It is about 12 kilometers in length and one kilometer in breadth. It is fed by *Choa Nala* (Jhambowali Choi), and is full of water during rainy season. It contains fewer water levels in summer months.

Besides, the district has a large number of water ponds. These are typical of each village settlement. Invariably, the water of village settlements drains into them. These are the heavens of water for buffaloes who bath in them, especially during summer months for evading scorching heat. Some of these ponds get dried in summer. They are full of water during the monsoon period. These act as breading grounds for mosquitoes.

3.6   GROUNDWATER SCENARIO

The ground water occurs in alluvium formations comprising fine to coarse sand, which forms the potential aquifers. In the shallow aquifers (up to 50m), groundwater occurs under unconfined/water-table conditions, whereas in deeper aquifers, semi-confined/confined conditions exist. This aquifer is tapped by the hand pumps and shallow tube wells, which are widely used for domestic purposes. The depth to water level ranges from 4.43 to 20.62 m bgl during pre-monsoon period and 6.99 to 24.28 m bgl during post monsoon period. The seasonal fluctuation varies from 0.03 to (-) 3.66 m in the area. The long-term water level trends indicate average fall of 0.50 m/year. The long term water level trend is also showing little rise being 0.24 m/year around majauli, which is insignificant with respect to entire area.

The elevation of the water table in the district varies from 230m to 300m above mean sea level. The highest elevation is in the northeastern part and the lowest in the southwestern part and reflects the topographic gradients. The hydraulic gradient in the northern eastern part is steep, whereas, in the southwestern part, it is gentle. The overall flow of ground water is from northeast to south-west direction.

Table 3.1: Groundwater Resources Availability, Utilization and Stage of Development of Patiala District in Comparison to Punjab (March 2004)

	Annual Replenishable Groundwater Resources (bcm)	Natural Discharge during Non-Monsoon (bcm)	Net Groundwater Availability (bcm)	Annual Groundwater Draft (bcm)	Groundwater Availability for Future (bcm)	Groundwater development %
Patiala	1.81	0.18	1.63	2.69	-1.07	165
Punjab	23.78	2.33	21.44	31.16	-9.89	145
(Source: Central Groundwater Board, 2006)						

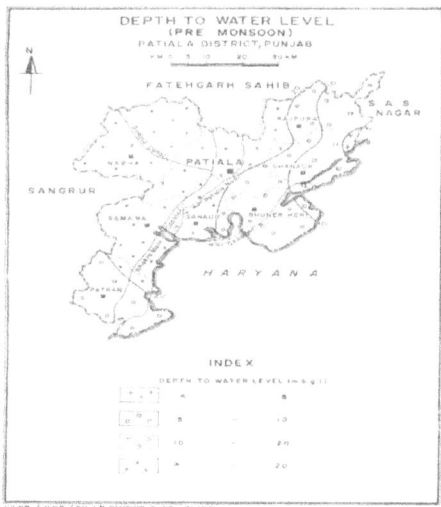

Figure 3.4: Patiala-Depth to Water Level; Pre-Monsoon

Figure 3.5: Patiala-Depth to Water Level; Post-Monsoon

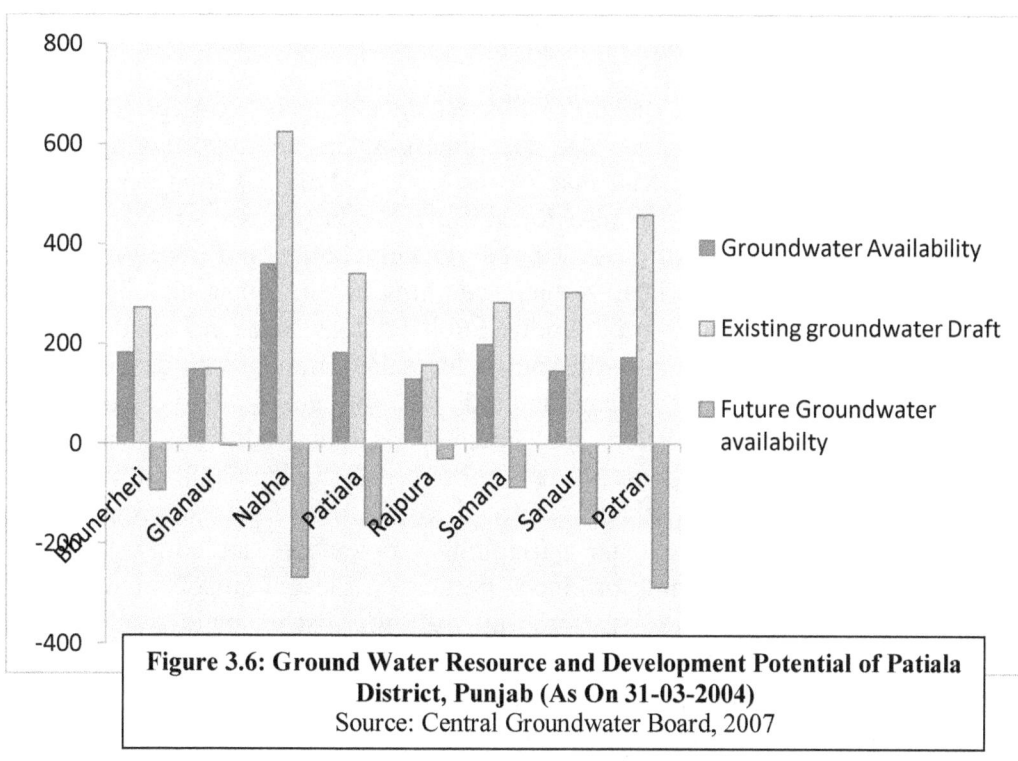

Figure 3.6: Ground Water Resource and Development Potential of Patiala District, Punjab (As On 31-03-2004)
Source: Central Groundwater Board, 2007

Figure 3.7: Ground Water Resource and Development Potential of Patiala District, Punjab (As On 31-03-2004)

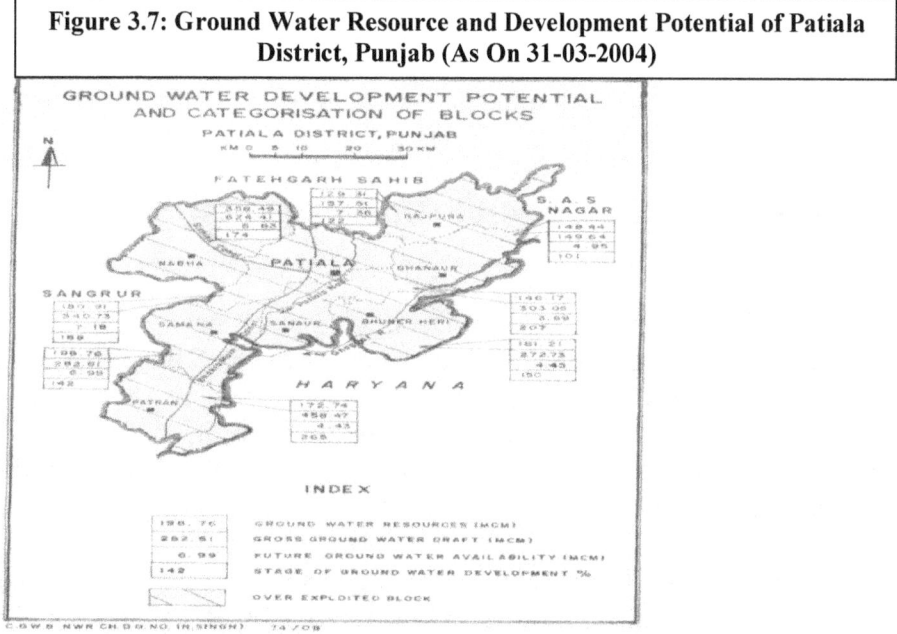

The net ground water resource of Patiala district have been estimated to be 1516.03mcm and the gross ground water draft of the district is 2589.45mcm leaving behind a shortfall of (-) 1087.69mcm. The stage of ground water development in the district is 169%. The block-wise groundwater resource potential in the district is available in the Figure No. 3.4 to Figure No. 3.7.

3.7   INDUSTRIAL DEVELOPMENT

Though the people of Patiala District depend mainly on agriculture for earning their livelihood, it is also fast emerging as an important industrial growth centre on the industrial map of the state. Patiala has long been known for its gota and zari work as also for silk azarbands (trouser strings) , daryai (silk cloth) due to the patronage of the princely house and gentry susi and durries were made at Patiala, Nabha and Amloh, and brass and bell metal utensils at Patiala. Sanour was famous for hand fans made out of date palm leaves, patiala and Nabha are still known for nala, parandas and desi juti. The jails at Patiala and Nabha manufactured and still continue to manufacture durries, niwar and ban of very high quality. Besides traditional goods, high quality and sophisticated items are now produced including small cutting tools, industrial cables, power cables, electrical goods, milkfood, vanaspati ghee, biscuits, horlicks, bicycles and agriculture implements including harvester combines and threshers, milk products, pesticides, diesel component workshop etc. The industrial units are scattered all over the district mainly at Rajpura, Dera bassi, Patiala, Samana and Nabha. There are large and medium industrial units located at Rajpura producing vanaspati ghee, power cables, bicycles and bicycle components. Among the small scale industry in the district are those producing agriculture implements, rice shellers, cutting tools, electrical goods and bakeries.

There are Industrial Focal Points at Patiala, Rajpura and Nabha and three Industrial Estates at Rajpura, Patiala and Banur.

3.8   GHAGGAR RIVER and INDUSTRIAL POLLUTION

The Ghaggar originates from the feet of Shivalik hills and runs towards North-East of Kalka. It enters Punjab in Dera Bassi block near village Kakrali and passes through Patiala, Sangrur and Bathinda districts of Punjab. It leaves Punjab near village Moonak after covering a distance of about 180 km in the Punjab territory. High values of Mercury, heavy metals and pesticides have been reported in water in 2003 at Dhakansu nallah (PSCST, 2005). Further, values of most metals are high in sediments. With regard to pesticides, high values have been reported at Khanauri, both, in water and sediments in certain stretches of the Ghaggar River which are being monitored by CPCB. Punjab Pollution Control Board identified the following sources of pollution in river Ghaggar which are listed as under:

3.8.1   Industrial Sources

The following industries were found by PPCB, discharging their wastewater into river Ghaggar:

- M/s Chandigarh Distillers & Bottlers Ltd., Village Banur was found discharging its effluent occasionally into river Ghaggar through Dakansu Nallah.
- M/s Surya Pharmaceuticals Ltd., Village Banur was found discharging its effluent occasionally into river Ghaggar through Dakansu Nallah.
- M/s PSIEC, Focal Point, Dera Bassi, where 10 out of 49 industrial units are water polluting and disposing wastes into river Ghaggar.
- M/s Milkfood Ltd., Bahadurgarh disposing wastes into drain which leads to river Ghaggar.

3.8.2 Domestic Sources

The following Local bodies were identified by PPCB, discharging their wastewater into river Ghaggar:

Zirakpur, Dera Bassi, Mohali, Bassi Pathana, Sirhind, Mandi Gobindgarh, Patiala, Rajpura, Nabha, Sanour, Banur, Khanauri, Sardoolgarh, Moonak, Vil. Ratnaheri, Vil. Mandvi, Vil. Makrour, Vil. Phulad, Vil. Bhunder.

The drinking water supply is mainly through ground water in the district. The short fall in water supply to towns, cities and villages is met with the installation of hand pumps by public individually as spot and convenient source of water. The shallow tube wells tap unconfined aquifer and depth varies from 20 to 70m.The tube wells constructed by the municipal corporation and other agencies have been constructed tapping deeper aquifer down to 100m. The shallow tube-wells irrigate about 2670 sq.km. area in the district. Most of these shallow tube-wells are cavity type and either run by diesel engines or electric motors. The discharge of these shallow tube wells/cavity wells range 600 – 1000 lpm.

Table 3.2: Coverage of Habitations with Water Supply in Patiala district (Mar-2006)							
Type of Habitations	Fully covered		Partially covered		Not covered		Total
	Nos.	%	Nos.	%	Nos.	%	Nos.
Main habitations	773	73.06	251	23.72	34	3.21	1058
Other habitations	57	45.24	27	21.43	42	33.34	126
Total habitations	830	70.1	278	23.48	76	6.42	1184
(Source: Department of Water Supply and Sanitation, Punjab, 2006)							

Chapter-4

DRINKING WATER QUALITY PARAMETERS

Groundwater represents an important source of drinking water but its quality is currently threatened by over-abstraction and microbiological and physicochemical contamination (Reid et al., 2003; Pedley and Howard, 1997).

4.1   WATER and its CONTAMINATION

Being universal solvent, water dissolves almost all, including toxic and hazardous substances, producing polluted water. Many physicochemical parameters may be of interest for water quality assessment like, temperature, colour, odour, pH, turbidity, salinity, hardness, nitrates, phosphates and certain trace elements. Temperature has a marked influence on the chemical and biochemical reactions that occur in water bodies. High temperature increases toxicity of many substances and sensitivity of living organisms (Dojlido and Best, 1993).   Low pH accelerates the corrosion of metals. Presence of suspended matter in water is termed as turbidity which absorbs and scatters the incoming light. WHO set the maximum permissible limit of turbidity at 5 NTU (Nephalometric Turbidity Unit). Lawson and Mason (2001) studied a strong correlation between metals like lead, arsenic, cadmium etc. and suspended particulate concentration. Hardness in water is caused by dissolved calcium, and to a lesser extent, by magnesium. Acceptable hardness ranges between 100 and 200 mg $CaCO_3$ per litre. Hardness above 200 can result in scale deposition and below 100 leads to low buffering capacity. Intake of very soft waters may have an adverse effect on mineral balance and can cause cardiovascular diseases, rectal and esophageal cancer and even mortalities (Sauvant and Pepin, 2000; WHO, 1996; Young et al., 1999a; 1999b; Dojlido and Best, 1993). Nickel allergy and hand eczema due to exposure to nickel, kidney damage due to cadmium, motor-neuron disease and poor verbal and visual memory due to high manganese have been reported (Vahter et al., 2002; Woolf et al., 2002; Iwami et al., 1994). Acute lead leads to tiredness, lassitude, slight abdominal discomfort, irritability, anaemia, neuro-physiological effects and behavioral changes in children. Zinc toxicity symptoms include vomiting, dehydration, electrolyte imbalance, abdominal pain, nausea, lethargy, dizziness and lack of muscular coordination. Acute renal failure caused by zinc chloride has been reported (WHO, 1991). Nitrates in groundwater are often associated by farming, fertilizers, pesticides or poor sanitary activities (Jacinthe et al., 2000; Neal et al., 2000; Nolan and Stoner, 2000; Huang et al., 1994). WHO guidelines for nitrates in drinking water is established to prevent methaemoglobinaemia (blue babies) which is lethal in babies and can be potentially hazardous with health risks for considerable groups of people who live near shallow or dug wells or springs where nitrate concentration is high (Squillace et al., 2002; Lehloesa and Muyima, 2000). The condition can progress rapidly to coma and death if it is not recognized and treated appropriately (Knobeloch et al., 2000). Trace elements of interest in drinking water of non-industrial areas are cadmium and lead. Cadmium accumulates in kidneys and damages it. Growth of algae, the blue-green algae in particular, can produce toxins as byproduct of photosynthesis in water supply reservoirs (WHO, 1996; Dojlido and Best, 1993; Zakowa, 1993). Lehloesa and Muyima (2000) have studied the adverse effects of lead. Lead is toxic to both the central and peripheral nervous system (WHO, 1993). Clement et al. (2000) claims to estimate the exposure of a population to lead contaminated by drinking water supply. When water is run from a tap, the first fraction of water collected carries the highest lead concentration, since the water was standing in the lead pipe. Well or storage tanks also have high levels of lead (Clement et al., 2000; Wyatt et al., 1998) in addition to tap waters and springs (Leroyer et al., 2000; Chu et al., 1998). Srikanth et al. (2002), Shivashankara et al. (2000), and many others have studied the ill effects of fluoride on teeth and bones. The major observed health effect of sulphate is its laxative action (WHO, 1996). Chloride concentration in excess of about 250 mg/L can give rise to detectable taste to water (Kim et al., 2002).

Microorganisms' threat to safety of drinking water is a growing peril even in industrialized nations that have long regarded themselves as immune to wide spread water borne illness and carriers so common in developing countries (Young, 1996; WHO, 1991). Microbial pathogens including bacteria (E. coli, Salmonella, Shigella, Vibrio Cholerae, Campylobactor, Yersinia, Klebseilla etc.), viruses (Hepatitis Virus, Rotavirus etc.), Cyanobacteria (toxins) and Protozoa (Giardia, Entamoeba, Cryptosporidium etc.) are major health risks associated with water and wastewater (Eynard et al., 2000; Maurer and Stuerchler, 2000; Szewzyk et al., 2000; Toze, 1999; Pathak and Gopal, 1994). In the developing world, drinking water is an important route for transmission of diarrheal disease, especially in rainy season (Dangendorf et al., 2002; Thevos et al., 2000; Musa et al., 1999). Ice used for human consumption can also be contaminated with pathogenic microorganisms and may become a vehicle for human infection (Falcao et al., 2002). Microbiologically contaminated drinking water is a cause of community acquired infection, especially water borne nosocomial pneumonias caused by Pseudomonas aeruginosa (Anaissie et al., 2002; Dangendorf et al., 2002). The number of immune-compromised patients who are at risk of developing infectious diseases from exposure to contaminated water aerosolized from dental units connected to municipal water supplies will continue to increase (Plamondon and Mills, 2000). Rainfall is major factor affecting vertical and horizontal movements of bacteria in soil. Surface runoff carries bacteria significant distances downstream causing serious threats to surface and ground water. Soil texture plays an important role; E. coli survives in semi-arid areas for a long time and increases potential of contamination

(Ashour and Hung, 2000). Another threat mentioned by Power and Nagy (1999) is bacterial re-growth present within the system and certain parameters such as turbidity and distance from the initial treatment point, correlated with the presence of high bacterial numbers. Zachcus et al. (2001) confirmed further that soft pipeline deposits were found to be the key site for microbial growth in the distribution networks.

It is impossible to test the water supply for all pathogens related to water borne diseases because of the complexity of the testing, the time and cost related to it (Lee and Kim, 2002; Toze, 1999). It is therefore preferable to use indicator systems which are able to index the presence of pathogens in water and wastewaters. The presence of pathogens is usually accompanied by the presence of classic indicators of contamination such as E. coli, Enterococci and other aerobic bacteria (Schaffter and Parriaux, 2002; Charriere et al., 1994). The microbiological drinking water guidelines aim at ensuring both the protection of human health and evaluation of the treatment efficacy.

Water quality is the combination of properties of water that are manifested in relation to human, other living creatures, items and substances. Due to the diversity of the natural forms of the existence of water and diversity of forms of water using by humans (biological and technical), the vast multiplicity of water properties was explored (physical, chemical, biological, and technological). Every specific sort of water requires a special method of quality analysis. However, in practice, general methods of analysis of large groups of water are used: natural water for potable water supply, industrial for use in the production processes, and waste water for discharge into water basins or for further use. Depending on the field of application of the water, the specific requirements are presented for its quality along predetermined parameters.

Chemicals that are toxic and might be found in drinking water may cause either acute or chronic health effects. An acute effect usually follows a large dose of a chemical and occurs almost immediately. Examples of acute health effects are nausea, lung irritation, skin rash, vomiting, dizziness, and, in the extreme, death. The levels of chemicals in drinking water, however, are rarely high enough to cause acute health effects. They are more likely to cause chronic health effects, effects that occur after exposure to small amounts of a chemical over a long period. Examples of chronic health effects include cancer, birth defects, organ damage, disorders of the nervous system, and damage to the immune system. The possible health effects of a contaminant in drinking water differ widely, depending on whether a person consumes the water over a long period, briefly, or intermittently.

4.2 SETTING of MAXIMUM CONTAMINANT LEVELS

Standards set by National Drinking Water Authority are called Maximum Contaminant Levels (MCL). MCL is the highest amount of a specific contaminant allowed in the water delivered to any consumer. An MCL may be expressed in milligrams per liter (mg/L), which is the same for the purposes of water quality analysis as parts per million (ppm) or as micrograms/liter (μg/L) equivalent to parts per billion (ppb) or one thousand micrograms per liter (1000μg/L) equivalent to one milligram per liter (1mg/L). MCL have been set by the USEPA and the BIS (in India) to provide a margin of safety to protect the public health.

Impurities in drinking water that are regulated and have an adverse health impact are grouped into six categories: inorganic chemical contaminants, volatile organic chemical contaminants, synthetic organic chemical contaminants, microbiological contaminants, radiological contaminants, and disinfection by-products. The process of setting MCL for drinking water contaminants is based on three criteria:

➢ The contaminant causing adverse health effects;
➢ Instruments available to detect the contaminant in drinking water; and
➢ The contaminant is known to occur in drinking water.

Experts use available chronic and sub-chronic animal studies and human clinical or epidemiological data to estimate the concentration of a drinking water contaminant that may be toxic and the concentrations, if any, which may cause no adverse effects. Acceptable Daily Intake called the Reference Dose is used to establish a Maximum Contaminant Level Goal for a contaminant. The standard-setting process was specified in the 1983 amendments to the New Jersey Safe Drinking Water Act (A-280) and is in some ways different from the federal process for these chemicals. In 1987, USEPA published MCL for eight volatile organic chemicals. Setting drinking water standards is an imperfect process, rarely based on conclusive human evidence. In addition, regulatory decisions frequently incorporate economic, political, and social considerations. The judgment of safety or what is an acceptable level of risk in particular circumstances is a matter in which society as a whole has a role to play.

4.3 CONSUMERS' AWARENESS

In assessing the quality of drinking-water, consumers rely principally upon their senses. Microbial, chemical and physical water constituents may affect the appearance, odour or taste of the water and the consumers generally evaluate the quality and acceptability of the water on the basis of these criteria. The quality of drinking-water may be controlled through a combination of protection of water sources, control of treatment processes and management of the distribution and handling of the water. Effective communication to increase community awareness and knowledge of drinking-water quality issues and the various areas of responsibility helps consumers to understand and contribute to decisions regarding safe and pure drinking water. The acceptability of drinking-water to consumers is subjective and can be influenced by many different constituents. The concentration at which constituents are objectionable to consumers is variable and dependent on individual and local factors, including the quality of the water to which the community is accustomed and a variety of social, environmental and cultural considerations.

4.4   SAMPLING OF THE DRINKING WATER

Drinking water samples were collected from villages of all the eight i.e. Bhunerheri, Ghanaur, Nabha, Patran, Patiala, Rajpura, Samana and Sanour blocks of Patiala district. Convenient non-probability sampling method was employed to choose the water samples. A total of 200 samples were analyzed for different quality parameters like Temperature, pH, Electrical Conductivity, Total dissolved solids, Total Suspended Solids, Chloride, Total hardness, Calcium, Magnesium, Total Alkalinity, Fluoride, Nitrate and Microbiological assessment.

The samples were collected in pre-cleaned, sterilized, polyethylene bottles of one litre capacity. It was ensured every time that bottle satisfies the following requirements: Free from contamination, resistant to any internal pressure and do not affect water characteristics. The water was left to run from the sampling source for 4-6 minutes and then collected in the bottles. For sampling of the water, cluster sampling method was employed. Five samples of one litre each were collected from the village, considering distance and source and then samples were mixed and one litre of that was used for analysis. It was needed keeping in mind the village people habits of drinking water from various sources.

Sampling was done in the months of April 2009 for Patiala and Patran block, September 2009 for Bhunerheri and Sanour block, April 2010 for Nabha, Rajpura and Samana block and September 2010 for Ghanaur block. Hand pumps or tube wells generally extract water from the top water bearing strata. The depth of the hand-pumps installed by rural people was less than 100feet, private tube-wells depth was approx. 300feet and the depth of tube-wells installed by government to supply drinking water in some of the rural areas was approx. 700-1100feet.

The sampling bottles were numbered as given below:

4.4.1 BHUNERHERI BLOCK:

A total of 30 water samples from the villages in this block were studied viz.

Abdulpur (Bh-01),	Bangran (Bh-02),	Behru (Bh-03),
Bhankher (Bh-04),	Bhasmra (Bh-05),	Chirwa (Bh-06),
Chuhat (Bh-07),	Daulatpur (Bh-08),	Faridpur (Bh-09),
Ghuram (Bh-10),	Goherpur (Bh-11),	Guthmera (Bh-12),
Hajipur (Bh-13),	Harana (Bh-14),	Jodhpur (Bh-15),
Julkan (Bh-16),	Kapoori (Bh-17),	Mahman (Bh-18),
Mahru (Bh-19),	Mandi (Bh-20),	Mauran (Bh-21),
Mulgarh (Bh-22),	Panjeta (Bh-23),	Rahgarh (Bh-24),
Rattakhera (Bh-25),	Roshanpur (Bh-26),	Sadhipur (Bh-27),
Sarkara (Bh-28),	Shekhupur (Bh-29)	Surastigarh (Bh-30).

4.4.2 GHANAUR BLOCK:

For this block, a total of 30 water samples from the villages were studied viz.

Ajror (Gh-01),	Alamdipur (Gh-02),	Bapror (Gh-03),
Bhat Majra (Gh-04),	Bhuri Majra (Gh-05),	Bibi Pur (Gh-06),
Chamaru (Gh-07),	Chappar (Gh-08),	Dharian (Gh-09),
Hari Majra (Gh-10),	Harpalpur (Gh-11),	Hassanpur (Gh-12),
Jharwan (Gh-13),	Kabulpur (Gh-14),	Kamalpur (Gh-15),
Kapuri (Gh-16),	Kole Majra (Gh-17),	Lachhru Kalan (Gh-18),
Manjoli (Gh-19),	Mirzapur (Gh-20),	Mughal Majra (Gh-21),
Nanhera (Gh-22),	Nanheri (Gh-23),	Nogawan (Gh-24),
Noshehra (Gh-25),	Rasulpur (Gh-26),	Rurka (Gh-27),
Rurki (Gh-28),	Tepla (Gh-29)	Uski (Gh-30).

4.4.3 NABHA BLOCK:

A total of 30 water samples from the villages were studied in this block viz.

Alampur (Nb-01),	Alhoran (Nb-02),	Alipur (Nb-03),
Bazidri (Nb-04),	Chehal (Nb-05),	Dangera (Nb-06),
Daroki (Nb-07),	Doda (Nb-08),	Faizgarh (Nb-09),
Ghanurki (Nb-10),	Gunika (Nb-11),	Gurditpura (Nb-12),
Ichhewal (Nb-13),	Jindalpur (Nb-14),	Kala Majra (Nb-15),
Khizarpur (Nb-16),	Kotli (Nb-17),	Kularan (Nb-18),
Lope (Nb-19),	Malehwal (Nb-20),	Malkon (Nb-21),
Maugo (Nb-22),	Nanowal (Nb-23),	Paharpur (Nb-24),
Pedan (Nb-25),	Raisal (Nb-26),	Sauja (Nb-27),
Shahpur (Nb-28),	Srinagar (Nb-29)	Thuhi (Nb-30).

4.4.4 PATRAN BLOCK:

A total of 20 water samples from the villages in this block were studied viz.

Arneto (Pr-01),   Bhkraha (Pr-02),   Bhootgarh (Pr-03),
Chungara (Pr-04),   Dhabi Gujran (Pr-05), Dhuhar (Pr-06),
Duttal (Pr-07),   Galoli (Pr-08),   Hamjeri (Pr-09),
Jaikher (Pr-10),   Jeonpur (Pr-11),   Jogewal (Pr-12),
Matauli (Pr-13),   Naiwala (Pr-14),   Nial (Pr-15),
Noorpura (Pr-16),   Salewala (Pr-17),   Seona (Pr-18),
Shabilpur (Pr-19)   Sher Garh (Pr-20).

4.4.5 PATIALA BLOCK:

A total of 20 water samples from the villages in this block were studied viz.

Alampur (Pt-01),   Assa Majra (Pt-02),   Bibipur (Pt-03),
Boharpur (Pt-04),   Chahohona (pt-05),   Challela (Pt-06),
Chamarheri (Pt-07),   Fatehpur (Pt-08),   Gajju Majra (Pt-09),
Mirzapur (Pt-10),   Mithu Majra (Pt-11),   Muradpur (Pt-12),
Phalauli (Pt-13),   Raipur (Pt-14),   Raj Garh (Pt-15),
Rithkheri (Pt-16),   Shamspur (Pt-17),   Shankarpur (Pt-18),
Walipur (Pt-19)   Wazidpur (Pt-20).

4.4.6 RAJPURA BLOCK:

A total of 30 water samples from the villages in this block were studied viz.

Abdulpur (Rj-01),   Alampur (Rj-02),   Aluna (Rj-03),
Basman (Rj-04),   Bathli (Rj-05),   Bhadak (Rj-06),
Chamaru (Rj-07),   Chandaun (Rj-08),   Changera (Rj-09),
Chharwar (Rj-10),   Chhat (Rj-11),   Dhamoli (Rj-12),
Dhindsan (Rj-13),   Dhuman (Rj-14),   Faridpur (Rj-15),
Gado Majra (Rj-16),   Gajipur (Rj-17),   Hansla (Rj-18),
Harion (Rj-19),   Hulka (Rj-20),   Jansui (Rj-21),
Kale Majra (Rj-22),   Lolhan (Rj-23),   Mamoli (Rj-24),
Rangian (Rj-25),   Ranpur Kalan (Rj-26), Safdarpur (Rj-27),
Shamdo (Rj-28),   Thuha (Rj-29)   Uksi (Rj-30).

4.4.7 SAMANA BLOCK:

A total of 20 water samples from the villages in this block were studied viz.

Alampur (Sm-01),   Assarpur (Sm-02),   Bijalpur (Sm-03),

Chuhat (Sm-04),	Dullar (Sm-05),	Fatehpur (Sm-06),
Gajewas (Sm-07),	Gazipur (Sm-08),	Kularan (Sm-09),
Lalgarh (Sm-10),	Nagri (Sm-11),	Namada (Sm-12),
Nasupur (Sm-13),	Rajgarh (Sm-14),	Rajla (Sm-15),
Ratanheri (Sm-16),	Saidipur (Sm-17),	Sapar Heri (Sm-18),
Todarpur (Sm-19)	Ugoke (Sm-20).	

4.4.8 SANOUR BLOCK:

A total of 20 water samples from the villages in this block were studied viz.

Assarpur (Sr-01),	Bakshi Wala (Sr-02),	Balamgarh (Sr-03),
Bibipur (Sr-04),	Chaura (Sr-05),	Hamayunpur (Sr-06),
Jogipur (Sr-07),	Kakepur (Sr-08),	Kheri Gujran (Sr-09),
Kheri Jattan (Sr-10),	Lalina (Sr-11),	Langroi (Sr-12),
Paharipur (Sr-13),	Paki Khuhi (Sr-14),	Panjola (Sr-15),
Passiana (Sr-16),	Saphera (Sr-17),	Shadipur (Sr-18),
Sular (Sr-19)	Traine (Sr-20).	

4.5 PHYSICOCHEMICAL AND MICROBIOLOGICAL ANALYSIS METHODS

In the present study, 200 water samples were collected from eight blocks of Patiala district using convenient non-probability sampling method from the various villages of Patiala district. Sampling areas are mostly affected by agricultural and other anthropogenic activities besides effects of dumping industrial wastes into the land or nearby water source. For effective maintenance of water quality through appropriate control measures, continuous monitoring of large number of quality parameters is essential.

However it is very difficult and laborious task for regular monitoring of all the parameters even if adequate manpower and laboratory facilities are available (Bhandari and Nayal, 2008). The samples collected were analyzed for various water quality parameters like Temperature, pH, Electrical Conductivity, Total Dissolved Solids, Total Alkalinity, Total Suspended Solids, Total Hardness, Calcium, Magnesium, Fluoride, Chloride, Nitrate and Microbiological assessment following the ASTM (1972) and APHA (1989) standard methods. The analytical results were used as input for statistical analysis.

4.5.1 TASTE, ODOUR AND APPEARANCE

Taste and odour can originate from natural inorganic and organic chemical contaminants and biological sources or processes (e.g., aquatic microorganisms), from contamination by synthetic chemicals, from corrosion or as a result of water treatment (e.g., chlorination). Taste and odour may also develop during storage and distribution due to microbial activity. Taste and odour in drinking-water may be indicative of some form of pollution or of a malfunction during water treatment or distribution. It may therefore be an indication of the presence of potentially harmful substances.

4.5.2 TEMPERATURE

Cool water is generally more palatable than warm water, and temperature will impact on the acceptability of a number of other inorganic constituents and chemical contaminants that may affect taste. High water temperature enhances the growth of microorganisms and may increase taste, odour, colour and corrosion problems. In analysis of the physicochemical quality of pipe water samples, temperature is considered as a critical parameter. It has an impact on many reactions, including the rate of disinfectant decay and by-product formation (Volk *et al.*, 2002). As the water temperature increases the disinfectant demand and by product formation, nitrification, microbial activity, algal growth, taste and odour episodes, lead and copper solubility increases. Moreover, sand calcium carbonate precipitation also increases.

Thermometers were used for taking temperatures at the sampling sites.

4.5.3 pH ESTIMATION

pH is one of the most important operational parameters for water treatment such as disinfection or coagulation-flocculation and pH adjustment is a common practice in water treatment. Because, dissociation is poor at pH<6, at pH 6 to 8.5 a nearly complete dissociation of HClO occurs. Thus for disinfection with chlorine, control of pH is critical. As a consequence, an increasing pH of the potable water requires rising amounts of chlorine for the same disinfection efficacy. The microbial activity of chlorine is greatly reduced at high pH, probably because at an alkaline pH, the predominant species of chlorine is OCl⁻. Equilibrium concentrations of HClO and OCl depend on the pH of the water. If the pH of the water is high, chlorine is less effective in killing pathogens (USEPA, 1999).

The pH of the samples was analyzed at the site by using pH strips and Eutech Cybernetics Model pH Scan Meter. The pH meter was first calibrated with buffer solutions of pH 4.0, 7.0 and 9.2 and then pH of samples was determined.

4.5.4 ELECTRICAL CONDUCTIVITY

Conductivity reflects mineral salt contents of water and is an expression of its ability to conduct an electric current. As this property is related to the ionic content of the sample which is in turn a function of the dissolved (ionisable) solids concentration, the relevance of easily performed conductivity measurements is apparent. In itself conductivity is a property of little interest to a water analyst but it is an invaluable indicator of the range into which hardness and alkalinity values are likely to fall, and also of the order of the dissolved solids content of the water. The conductivity values can be used to find the Total Dissolved Solids.

The electrical conductivity (in dS/m) of the water samples was estimated by using a Eutech Cybernetics Model TDS Scan-I Meter. The EC meter was calibrated with standard KCl solution (0.1M). The standard KCI solution of 0.1 M was prepared by dissolving 0.7474 g of KCl (AR Grade) in 100 ml distilled water. The EC of standard solution was set at 12.88 dS at 25 °C. After calibration of the instrument, EC of the samples was recorded.

4.5.5 TOTAL DISSOLVED SOLIDS (TDS)

TDS were calculated indirectly making use of Electrical Conductivity (EC). To calculate the TDS, formula suggested by United States Salinity Laboratory Staff (1954) - Diagnosis and improvement of saline and alkaline salts, US Department of Agriculture, Hand Book 60, 160formula was used, which is given below:

$$TDS \ (mg/L) = 640 \times EC \ (in \ dS/m) \ (Hem, 1970)$$

4.5.6 TOTAL ALKALINITY

Requirements:

1. Standard sulphuric acid solution (N/50):
    2 ml of conc. H_2SO_4 was mixed with distilled water and made the volume 2 litres. H_2SO_4 solution was standardized with standard Na_2CO_3 solution using methyl orange indicator.
2. Phenolphthalein indicator.
3. Methyl orange indicator.

Procedure:

100 ml of water sample was taken in a flask and 2 drops of phenolphthalein indicator were added, if pink colour appeared, titrated it against

standardized H_2SO_4 solution until sample became colourless. The volume of acid used was recorded as 'A' ml. To the same solution, added 2-3 drops of methyl orange indicator and titrated it further with standard H_2SO_4 until colour changed from light yellow to pink. Again recorded the volume of acid consumed as 'B' ml. The experiment was repeated to get three concordant readings.

Calculations:

$$\text{Total alkalinity in mg/L (as } CaCO_3) = \frac{N \times V_2 \times 1000 \times EW}{V_1}$$

Where, N_2 = Normality of standardized H_2SO_4; V_2 = Volume (A+B) of acid used; V_1 = Volume of sample taken (100 ml); EW = Equivalent weight of $CaCO_3$ = 50

4.5.7 TOTAL SUSPENDED SOLIDS

Matter which is suspended in water consists of finely divided light solids which may never settle or do so only very slowly. The solids may consist of algal growths, washings from sandpits, quarries or mines. They reduce light penetration in water.

Procedure:

A well-mixed sample is filtered through a weighed (A) standard glass-fiber filter and the residue retained on the filter is dried to a constant weight (B) at 103 to 105°C. The increase in weight (B-A) of the filter represents the total suspended solids.

Calculations:

$$\text{Total Suspended Solids (mg/L)} = \frac{(B - A) \times 1000}{\text{Sample Volume (ml)}}$$

Where A = weight of filter (mg) and B = weight of filter + dried residue (mg).

4.5.8 TOTAL HARDNESS (TH) ESTIMATION (as $CaCO_3$)

Originally taken to be the capacity of water to destroy the lather of soap, hardness was determined formerly by titration with soap solution. Nowadays, the analysis comprises the determination of calcium and magnesium which are the main constituents of hardness. Although barium, strontium and iron can also contribute to hardness, their concentrations are normally so low in this context that they can be ignored. Thus, *total hardness* is taken to comprise the calcium and magnesium concentrations expressed as mg/L $CaCO_3$. The widespread abundance of these metals in rock formations leads often to very considerable hardness levels in surface and groundwater. Total hardness of water samples was determined by using EDTA titration method.

Requirements:

1. Standard EDTA solution (0.01M):
   3.723g of EDTA disodium salt was dissolved in double distilled water to make the volume one litre.
2. Ammonium buffer solution:
   16.9g of NH_4Cl was dissolved in 143 ml of NH_4OH and made the volume 250 ml using double distilled water.
3. Inhibitor solution:
   4.5g of hydroxylamine and hydrochloride ($NH_2OH.HCl$) was dissolved in absolute alcohol to make the solution 100 ml.
4. Erichrome Black-T (EBT) indicator:
   0.5g of EBT was dissolved in 100 ml of 80% ethyl alcohol.

Procedure:

25 ml of water sample was taken in a titration flask. To this, 2 ml of ammonia buffer solution, 1 ml of inhibitor and 3-5 drops of EBT indicator were added. It was titrated against EDTA solution till wine red colour changed to blue. The experiment was repeated to get three concordant readings.

Standardization of EDTA solution:

Prior to titration of EDTA with water samples, EDTA solution was standardized by using standard hard water ($CaCO_3$ solution) to find the actual molarity of EDTA solution.

Calculations:

$$\text{Total hardness (as } CaCO_3) = \frac{M_2 \times V_2 \times 1000 \times MW}{V_1} \text{ mg/L}$$

Where M_2 = Molarity of standardized EDTA solution; V_2 = Volume of EDTA solution used; V_1 = Volume of sample taken; M.W. = Molecular weight of $CaCO_3$ (100).

4.5.9 CALCIUM

Calcium content of water samples was estimated by using the complexometric titration method (EDTA).

Requirements:

1. Standard EDTA solution (0.01M):
   3.723g of EDTA disodium salt was dissolved in double distilled water to make the volume one litre. EDTA solution was standardized as given in total hardness estimation.
2. Standard NaOH solution (1 M):
   4g of NaOH was dissolved in double distilled water to make 100 ml solution.
3. Murexide Indicator (solid)

Procedure:

25 ml of water sample was taken in titration flask. To this, added 1 ml of 1M NaOH solution to raise the pH and a pinch of solid murexide. It was titrated with standard EDTA solution till the colour changed from pink to purple. The experiment was repeated to get three concordant readings.

Calculations:

$$Ca^{2+} (mg/L) = \frac{M_2 \times V_2 \times 1000 \times MW}{V_1}$$

Where, M_2 = Molarity of standardized EDTA solution; V_2 = Volume of EDTA solution used; M.W. = Molecular weight of $CaCO_3$ (100); V_1 = Volume of sample taken;

4.5.10 MAGNESIUM

Magnesium content of the water samples was calculated indirectly using Ca^{2+} and TH content of water samples by following formula:

$$Mg^{2+} (mg/L) = [\text{Total hardness} - (Ca^{2+} \times 2.5)] \times 0.243$$

4.5.11 FLUORIDE

BIS and WHO suggest desirable limit for Fluoride around 1.0 mg/L. At levels markedly over 1.5 mg/L an inverse effect occurs and mottling of teeth (or severe damage at gross levels) will arise. Fluoride was estimated by using Orion Selective Ion-electrode.

Principle:

The fluoride electrode is an ion-selective sensor. The key element in the fluoride electrode is the laser type doped lanthanum fluoride crystal across which a potential is established by fluoride solutions of different concentrations. The fluoride electrode can be used with a standard calomel reference electrode and almost any modern pH meter having an expanded milli-volt scale. The fluoride electrode measures the ion activity of fluoride in solution which depends on the total ionic strength of the solution. It does not respond to bond or complexed fluoride. Addition of a buffer solution of high total ionic strength (TISAB) overcomes these difficulties.

Reagents

1. Stock Fluoride solution:
    A standard solution of 1000 mg/L was prepared by dissolving 221.0mg anhydrous sodium fluoride in double distilled water and diluted to 100 ml.
2. Standard fluoride solution:
    Standard solution of 100 mg/L was prepared from stock fluoride solution.
3. Total ionic strength adjustment Buffer (TISAB):
    57 ml glacial acetic acid, 58gm NaCl and 4.0g 1, 2-cyclohexylenediamine tetracetic acid (CDTA disodium salt) was dissolved in 500 ml distilled water in a beaker. The beaker was placed in a cold water bath and 6M NaOH (about 125 ml) was added with stirring until pH was 5.4 ± 0.1 was achieved. The final volume was made one litre with double distilled water.

Procedure

A series of standard solutions with fluoride concentration 0.1, 1.0, 10.0 and 100 mg/L was prepared. 10 ml of the standard solution was taken in a beaker and diluted to 1:1 with TISAB. The electrodes were immersed in the solution and the developed potential was measured (in mV). A standard curve potential measurement (mV) vs. fluoride concentrations was prepared on two-cycle semi-logarithmic graph paper. Potential measurements were made for the collected water samples and fluoride concentration was calculated from the standard curve.

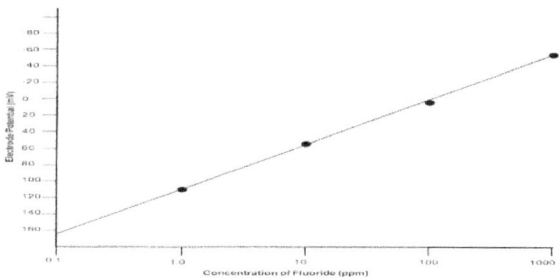

4.5.12 CHLORIDE

Chloride was determined by argentometric titration method:

Requirements:

1. Standard $AgNO_3$ solution (0.0141M):
    2.397 g of $AgNO_3$ was dissolved in one litre double distilled water.
2. 5% potassium chromate indicator solution:
    5 g K_2CrO_4 was dissolved in 100 ml distilled water.

Procedure:

25 ml water sample was taken in a titration flask and 3-4 drops of K_2CrO_4 indicator were added. It was titrated with standard $AgNO_3$ solution till yellow colour changed to light brick red. The titration was repeated to get three concordant readings.

Calculations:

$$Cl^- \ (mg/L) = \frac{M_2 \times V_2 \times 35.5 \times 1000}{V_1}$$

Where, M_2 = Molarity of standard $AgNO_3$ solution (0.141M); V_2 = Volume of $AgNO_3$ solution used in ml; V_1 = Volume of sample taken (25 ml).

4.5.13 NITRATE

Relatively little of the nitrate found in natural waters is of mineral origin, most coming from organic and inorganic sources, the former including waste discharges and the latter comprising chiefly artificial fertilizers. However, bacterial oxidation and fixing of nitrogen by plants can both produce nitrates. Interest is centered on nitrate concentrations for various reasons. Most importantly, high nitrate levels in waters to be used for drinking will render them hazardous to infants as they induce the "blue baby" syndrome (methaemoglobinaemia). The nitrate itself is not a direct toxicant but is a health hazard because of its conversion to nitrite which reacts with blood haemoglobin to cause methaemoglobinaemia.

Requirements:

- *Copper-cadmium granules:* Washed 25g mesh Cadmium granules with $6N$ HCI and then rinsed with water. Swirled Cd with 100 ml 2% $CuSO_4$ solution for 5 min or until blue color partially fades. Decanted and repeated with fresh $CuSO_4$ until a brown colloidal precipitate begin to develop. Then gently flushed with water to remove all precipitated Cu.
- *Ammonium chloride-EDTA solution:* Dissolved 13g NH_4Cl and 1.7g disodium ethylenediamine tetraacetate in 900 ml water. Adjusted the pH to 8.5 with conc. NH_4OH and dilute to 1 litre.
- *Dilute ammonium chloride-EDTA solution:* Diluted 300 ml NH_4Cl-EDTA solution to 500 ml with water.
- *Hydrochloric acid,* HCl, $6N$
- *Copper sulfate solution,* 2%: Dissolved 20g $CuSO_4.5H_2O$ in 500 ml water and dilute to 1 litre.
- *Color reagent:* To 800ml water added 100ml 85% phosphoric acid and 10g sulfanilamide. After dissolving sulfanilamide completely, added 1g N-(1-naphthyl)-ethylenediamine dihydrochloride. Mixed to dissolve, then diluted to 1 litre with water.
- *Nitrate free distilled water*
- *Spectrophotometer*

Procedure:

Inserted a glass wool plug into bottom of reduction column and filled it with water. Then sufficient Cu-Cd granules were added to produce a column of 18.5cm long. Maintained water level was above Cu-Cd granules to prevent entrapment of air. Then the column was washed with 200 ml dilute NH_4Cl-EDTA solution. Activated the column by passing through it at least 100ml of solution, composed of 25% (1.0mg NO_3^--N/L standard) and 75% (NH_4Cl-EDTA solution), at 7 to 10 ml/min.

pH of the sample water was adjusted to 7-9 by adding HCl or NaOH to ensure pH of 8.5 after adding NH_4Cl- EDTA solution. To 25ml sample, added 75ml NH_4Cl- EDTA solution and poured sample into column and collected at a rate of 7 to 10ml/min. Discarded first 25 ml. Collected the rest in flask. Then added 2ml color reagent to 50ml sample and measured absorbance at 543 nm against a distilled water-reagent blank.

Calculations:

Obtained a standard curve by plotting absorbance of standards against NO_3^--N concentration. Computed sample concentrations directly from the standard curve.

$$10 \text{ mg/L nitrate-nitrogen } (NO_3\text{-N}) = 44.3 \text{ mg/L nitrate } (NO_3\text{-})$$

4.5.14 TOTAL COLIFORMS

Coliform bacteria have been used to evaluate the general quality of water. Testing for coliform bacteria is faster and cheaper than testing for specific organisms and pathogens. So, these organisms are considered good indicators of the potential contamination of a water source. Coliform organisms are used as indicators of water pollution. The coliform organism is a very common rod-shaped bacterium, not thought of as disease causing to humans. Because pathogenic bacteria in wastes and polluted waters are usually much lower in numbers and much harder to isolate and identify than coliforms, which are usually in high numbers in polluted water, total coliforms is used as a general indicator of potential contamination with pathogenic organisms.

Reagents and Culture Medium:

(A) Lauryl trytose broth:

- Tryptose   20g
- Lactose   5g
- Dipotassium hydrogen phosphate   2.75g
- Potassium dihydrogen phosphate   2.75g
- Sodium Chloride   5g
- Sodium lauryl sulphate   0.1g
- Bromocresol purple (to determine acid production) 0.01g
- Reagent grade water   1litre

(B) Brilliant green lactose bile broth:

- Peptone   10g
- Lactose   10g
- Oxgall   20g
- Brilliant green   0.0133g

- Reagent grade water   1 litre

Procedure:

Added dehydrated ingredients to water, mixed thoroughly, and heated to dissolve. Then sufficient medium (A) was dispensed in fermentation tubes. Fermentation tubes were arranged in rows of five tubes each in a test tube rack for making dilutions and measuring diluted sample volumes. Then inoculated tubes were incubated at 35°C. After 24 hours each tube was swirled gently and examined for growth, gas, and acidic reaction. Shades of yellow colour indicated positive presumptive reaction.

Gently shook the presumptive tubes showing gas or acidic growth to re-suspend the organisms. One loopfuls of the culture was transferred each to fermentation tubes containing brilliant green lactose (B) bile broth. The inoculated brilliant green lactose bile broth tubes were incubated at 35°C. Formation of gas in any amount in the inverted vial of the brilliant green lactose bile broth fermentation tube constitutes a positive confirmed phase. The MPN values were calculated from the number of positive brilliant green lactose bile tubes. A magnifying glass was used to determine colony counts on the filter papers. Red and blue colonies combined indicated the sample had coliforms.

MPN/100ml = N × 100 /V

Where N is the no. of colonies counted and V the sample volume in ml.

Figure 4.1: Various types of Coliforms available in Drinking Water
(Source: http://www.freedrinkingwater.com/water_quality)

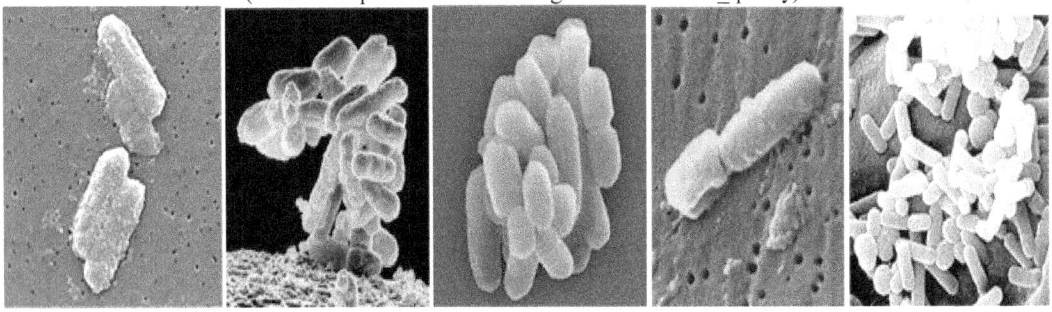

4.6   STATISTICAL INTERPRETATIONS

Water quality parameters were analyzed for the samples collected and statistical interpretations were carried out. All Mathematical and Statistical Calculations like Mean, Standard Deviation, Variance and Correlation Coefficients were implemented using Microsoft Office Excel 2007.

Chapter-5

DRINKING WATER QUALITY ANALYSIS OF DISTRICT PATIALA

Different experiments were conducted to achieve the objectives. For the sake of simplicity and convenience, the observations and results have been studied as given below.

- ✓ Block-wise Study
- ✓ Group and Cluster-wise Study
- ✓ District-wise Study

5.1   BLOCK-WISE STUDY

- ❖ Water quality monitoring in Bhunerheri Block,
- ❖ Water quality monitoring in Ghanaur Block,
- ❖ Water quality monitoring in Nabha Block,
- ❖ Water quality monitoring in Patran Block,
- ❖ Water quality monitoring in Patiala Block,
- ❖ Water quality monitoring in Rajpura Block,
- ❖ Water quality monitoring in Samana Block, and
- ❖ Water quality monitoring in Sanour Block,

5.2   GROUP AND CLUSTER-WISE STUDY

The whole area of the study was divided into different groups and clusters:

Group A comprising the area where Ghaggar River passes which carries waste water,
Group B comprises the area where Patiale-wali Nadi passes which carries waste water,
Cluster One comprises where both Ghaggar and Patiale-wali Nadi affects, and
Cluster Second comprises other areas of the district Patiala.

- ❖ Group -A: Bhunerheri and Ghanaur blocks, and
- ❖ Group -B: Patiala and Sanour blocks.

    and

- ❖ Cluster -1: Bhunerheri, Ghanaur, Patiala, Sanour and Patran blocks, and
- ❖ Cluster -2: Nabha, Rajpura and Samana blocks.

Table 5.1: Drinking Water Specifications – Indian Scenario

Parameters	BIS Standards IS:10500:1991		Undesirable Effects outside the Desirable Limit
	Desirable Limit	Permissible Limit	
Taste, Odour	Unobjectionable and Agreeable		
pH	6.5 – 8.5	6.5 – 8.5	Affects mucous membrane; bitter taste; corrosion; affects aquatic life
Electrical Conductivity ds/m	-	-	-
Total Dissolved Solids mg/L	500	2000	Undesirable taste; gastro intestinal irritations; corrosion or incrustation
Total Alkalinity Mg/L	200	600	Taste becomes unpleasant, Boiled rice turns hard and yellowish, and Pulses become hard after boiling.
Total Suspended Solids mg/L	-	-	-
Total Hardness mg/L	300	600	Poor lathering with soap; deterioration of the quality of clothes; scale forming; skin irritation; boiled meat and food become poor in quality
Calcium mg/L	75	200	Poor lathering and deterioration of the quality of clothes; incrustation in pipes; scale formation
Magnesium mg/L	30	75	Poor lathering and deterioration of clothes; with sulfate laxative
Fluoride mg/L	1.0	1.5	Dental and skeletal fluorosis; non-skeletal manifestations
Chloride mg/L	250	1000	Taste, corrosion and palatability are affected.
Nitrate mg/L	45	100	Blue baby disease (methemoglobineamia); algal growth
Total Coliforms MPN/100ml	Nil	Class A: 50 Class B: 500 Class C: 5000	Water should be used after disinfection and treatment to avoid water borne diseases due to pathogens.

(Source: Bureau of Indian Standards, New Delhi, 2003)

5.3   DISTRICT-WISE STUDY

❖ In this part the district as a whole is discussed.

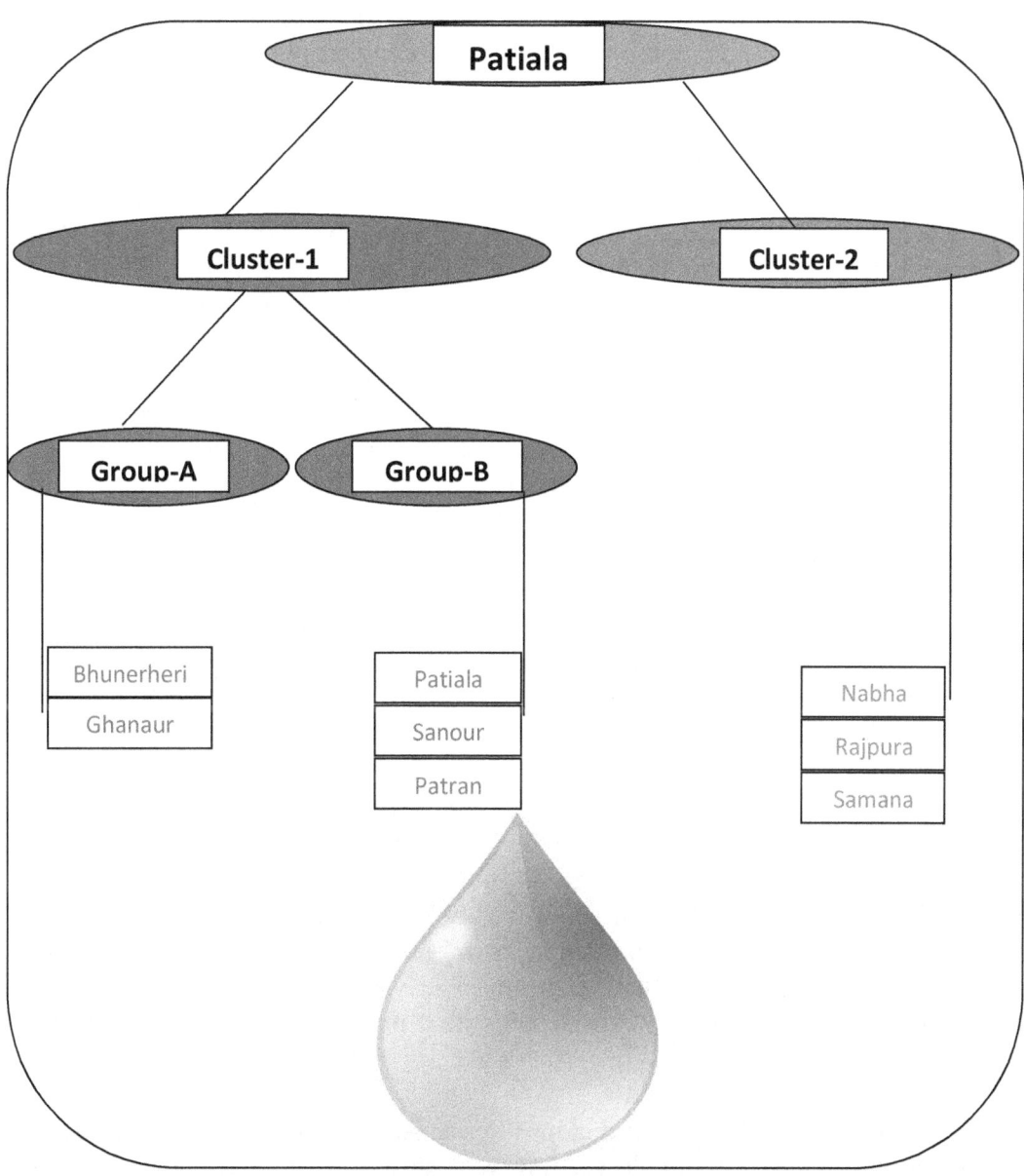

Figure 5.1: Scheme of Study for Drinking Water Quality Parameters

BLOCK-WISE STUDY

5.4   WATER QUALITY MONITORING IN BHUNERHERI BLOCK

Different drinking water parameters of the villages of this block as studied on the basis of monitoring water characteristics given in Table No. 5.2 and Table No.5.10 are presented below.

The pH indicates the intensity of acidity and alkalinity and measures the hydrogen ion concentration in water. In water, a small number of water (H_2O) molecules dissociate and form hydrogen (H^+) and hydroxyl (OH^-) ions. If the relative proportion of the hydrogen ions is greater than the hydroxyl ions, then the water is defined as being acidic. If the hydroxyl ions dominate, then the water is defined as being alkaline. The relative proportion of hydrogen and hydroxyl ions is measured on a negative logarithmic scale from 1 (acidic) to 14 (alkaline): 7 being neutral (Friedl *et al.,* 2004). pH is related and liable to change with temperature and pressure. In pure water, a decrease in pH of about 0.45 occurs with rise in temperature by 25°C (Kataria, 1995). Though pH has no direct effect on human health but all the biochemical reactions are sensitive to variation in pH. For most of the reactions as well as for human beings, pH value seven is considered best and ideal. The BIS permissible limits for drinking water are 6.5-8.5. In the present study regarding Bhunerheri block, the pH values in the drinking water samples of the study area ranged from 7.72 to 8.26 with the mean value of 8.00 and standard deviation 0.144 indicating slightly alkaline nature of water.

Electrical Conductivity signifies the amount of total dissolved salts, which in turn indicates the inorganic pollution load of the water. Electrical Conductivity ranges from 1.25 to 2.44mS with a mean value of 1.50mS and standard deviation 0.341. Total Dissolved Solids indicate the saline behavior of the water. BIS accepts 500 with a permissible limit of 2000 while WHO consider water up to 500 with a permissible value of 1500 safe. Bhunerheri block showed values from 801 to 1563 with a mean value 0f 987 and standard deviation 218.

Matter which is suspended in water consists of finely divided light solids which may never settle or do so only very slowly. The solids may consist of algal growths, washings from sandpits, quarries or mines. They reduce light penetration in water. In the present block the values ranged from 0.559 to 0.589 with a mean value of 0.576 and standard deviation 0.0077.

For its use in domestic sector for drinking, bathing and washing, Total Hardness of water plays a vital role. Externally hard water does not form lathers with soap leading to wastage of much quantity of soap and internally, if used for longer times, can be one of the causes of stone formation in the human body. Mainly the hardness of water is due to carbonates, bicarbonates, sulphates and chlorides of calcium and magnesium. The water of Bhunerheri block ranged hardness from 207 to 336mg/L of $CaCO_3$ with a mean value of 259 and standard deviation 44.78. Concentrations of calcium ranged from 30 to 65 with a mean value of 46.2, standard deviation 7.415 and concentrations of magnesium ranged from 22 to 64 with a mean value of 34.7, standard deviation 12.74.

The ability of the water to neutralize acids is known as Total Alkalinity. The constituents of alkalinity in natural systems include mainly carbonate, bicarbonate, hydroxide etc. These compounds result from dissolution of mineral substances present in the soil and atmosphere (Mittal and Verma, 1997). CO_3^{2-} and HCO_3^- may originate from microbial decomposition of organic matter also. Alkalinity is big problem for industries also, as if alkaline water is used in boilers for steam generation, it may lead to precipitation of sludges, deposition of scales and causes caustic embrittlement. The acceptable BIS limit of total alkalinity in drinking water is 200 mg/L with permissible limit 600. In present study Total Alkalinity ranged from 200 to 391 with a mean value of 247 and standard deviation 54.5.

Chloride contents in the water of Bhunerheri block ranged from 86 to 117 with a mean value of 99.7 and standard deviation 6.794.   Nitrate contents in the water of Bhunerheri block ranged from 2.013 to 3.927 with a mean value of 2.500 and standard deviation 0.548.

Coliforms are bacteria that are always present in the digestive tracts of animals, including humans, and are found in their wastes. They are also found in plant and soil material. The most basic test for bacterial contamination of a water supply is the test for total coliform bacteria. Total coliform counts give a general indication of the sanitary condition of a water supply. In the water of Bhunerheri block, total coliforms ranged from 81 to 157 with a mean value of 99 and standard deviation 21.910.

5.5   WATER QUALITY MONITORING IN GHANAUR BLOCK

Water monitoring data as observed regarding the villages of this block is given in Table No.5.3 and Table No.5.11. A total of 30 samples were studies from this clock. pH ranges from 7.69 to 8.54 with a mean value of 8.10 indicating alkaline nature of water. pH shows standard deviation 0.200.

Electrical Conductivity signifies the amount of total dissolved solids, which in turn indicates the inorganic pollution load of the water. Ghanaur block shows values ranges from 0.73 to 3.70 with a mean value of 1.30 and standard deviation 0.6311. Total dissolved solids indicate the saline behavior of water. World Health Organization considers 500 as desirable and 1500 as maximum permissible limit for drinking water. In Ghanaur block values ranged from 467 to 2367 with a mean value of 853 and standard deviation 403.879.

Total Suspended Solids ranged from 0.317 to 1.608 with a mean value of 0.580 and standard deviation 0.274. Total Alkalinity values of the water samples of this block ranged from 263 to 496 with a mean value of 319 and standard deviation 70.708. Total Hardness is from 159 to 504 with a mean value of 278 and standard deviation 101.669. Calcium ranged from 16 to 74 with a mean value of 28.8, standard deviation 12.918 and Magnesium ranged from 29 to 85 with a mean value of 50.0, standard deviation 17.362 indicating magnesium type soil as it has high concentration calcium.

Fluoride concentration vary from 0.212 to 0.563 with a mean value of 0.300, standard deviation 0.0974 and Chloride contents are 39 to 175, mean value 69.4 and standard deviation 30.208. Nitrate concentrations vary from 1.173 to 5.947 with a mean value of 2.100 and standard deviation 1.015. Total coliforms in this block also showed same type of trend from 47 to 238 with a mean value of 86 and standard deviation 40.591.

This block becomes in between Rajpura (rich in industrial works) and Ghaggar River which carries waste water from industrial areas of Rajpura, Mohali, Chandigarh, Haryana and Himachal Pradesh besides run-off of rainwater and agricultural effects.

5.6   WATER QUALITY MONITORING IN NABHA BLOCK

Water monitoring data regarding rural drinking water supplies in this block is available in Table no.5.4 and Table No.5.12. Nabha block showed pH ranges from 7.75 to 8.54 with a mean value of 8.09, standard deviation 0.185 indicating alkaline nature of water. Electrical Conductivity varies from 0.69 to 2.24 with a mean value of 1.12 and standard deviation 0.470. Total Dissolved Solids in this block ranges from 439 to 1431 with a mean value of 717 and standard deviation 300.523.

Total Alkalinity ranged from 218 to 492 with mean value of 293, standard deviation 80.258 and Total Hardness ranged from 220 to 516 with a mean value of 339 and standard deviation 112.876. Calcium with a mean value of 35.9 ranges from 22 to 72 and Magnesium showed mean value of 60.6 from 40 to 89 indicating magnesium high soils. Standard deviations for calcium and magnesium are 15.026 and 18.898, respectively. The mean values for Nitrate and Total Suspended Solids are 1.802 and 0.560 with the range 1.103 to 3.595 and 0.343 to 1.118 and standard deviations 0.755 and 0.235 respectively.

Fluoride ranges from 0.293 to 0.655 with a mean value of 0.413, standard deviation 0.118 and Total coliforms ranged from 22 to 72 with a mean value of 36, standard deviation 15.102.

5.7   WATER QUALITY MONITORING IN RAJPURA BLOCK

Water monitoring data regarding the villages of this block is presented in Table No.5.5 and Table No.5.13. A total of 30 samples were studied in this block. The water sample studied in this block show pH ranged from 6.98 to 8.17 with a mean value of 7.64, slightly alkaline nature of water and standard deviation 0.421. Total Dissolved Solids ranged from 376 to 1476 and Electrical Conductivity from 0.59 to 2.31 with mean values 726, 1.14 and standard deviation 293.834, 0.459 respectively. Total Suspended Solids varies from 0.294 to 1.153 with a mean value of 0.567 and standard deviation 0.230.

Total hardness, Calcium and Magnesium contents ranges from 188 to 738, 50 to 197 and 15 to 60 with the mean values 363, 96.8 and 29.4, respectively. Standard deviation for these parameters is 146.217, 39.178 and 11.900. Total Alkalinity value was found to be 277 from 139 to 579 and standard deviation 113.685.

Fluoride contents were 0.147 ranging from 0.076 to 0.299, Chloride contents were 110.9 from 56 to 232 and nitrate was 7.407 from 3.837 to 15.061. Standard deviation for these parameters was found to be 0.0595, 45.474 and 8.990 respectively. Total Coliforms ranged from 102 to 402 with a mean value of 198 and standard deviation 79.955.

Rajpura block is adjoining Mohali and both are industrial points. Possibly industrial/domestic wastes and agricultural effects are the causes of behavior shown by this block.

5.8   WATER QUALITY MONITORING IN PATIALA BLOCK

Water monitoring data regarding this block is presented in Table No.5.6 and Table No.5.14. pH ranged from 6.95 to 8.42 in this block with a mean value of 7.80 and standard deviation 0.4338. Electrical Conductivity ranged from 0.74 to 2.57 with a mean value of 1.42 and standard deviation 0.5381. Total Dissolved Solids ranged from 473 to 1643 with a mean value of 912 and Total Suspended Solids ranged from 0.370 to 1.284 with a mean value 0.712. Standard deviations for these Parameters were 344.366 and 0.2690 respectively.

Total Hardness value at 456, Calcium 121.5, Magnesium 36.9 and Total Alkalinity 349 showed range from 237 to 822, 63 to 219, 19 to 67 and 195 to 571 with standard deviation 172.183, 45.915, 13.947 and 114.671 respectively.

Fluoride found to be 0.369 from 0.192 to 0.665, standard deviation 0.1395, Chloride at 139.6 from 78 to 228, standard deviation 45.869 and Nitrate showed range from 1.587 to 5.513 with a mean value of 3.057, standard deviation 1.1556.

The values range from 42 to 147 for Total Coliforms with a mean at 82 and standard deviation 30.8157.

5.9   WATER QUALITY MONITORING IN PATRAN BLOCK

Water monitoring data is available in Table No.5.7 and Table No.5.15. The values in Patran block ranged for pH from 7.86 to 8.36 with a mean of 8.05, standard deviation 0.1605. Electrical Conductivity has the values from 0.75 to 2.46 with a mean of 1.50 and standard deviation of 0.5483. The values range from 482 to 1572 for Total Dissolved Solids with a mean 962 and standard deviation 350.9249.

Total Hardness values are from 241 to 786, mean value 481, standard deviation 175.4624, Calcium ranged from 64 to 210 with a mean value 128.3, standard deviation 46.790, Magnesium ranged from 20 to 64 with a mean value 38.96, standard deviation 14.2125 and Total Alkalinity values ranged from 192 to 623 with mean 387 and standard deviation of 140.6072.

Total Suspended Solids ranges for this block from 0.377 to 1.228 with a mean value of 0.752 and standard deviation 0.2742.

Fluoride and Chloride values for this block are 0.390 ranging from 0.195 to 0.637, standard deviation 0.14212 and 63.77 ranging from 31 to 106, standard deviation 23.443. Nitrate contents are at 4.859 from 2.434 to 7.939, standard deviation 1.7723.

The values showed by this block for Total Coliforms are 65 to 212 with a mean of 130 and standard deviation of 47.26262.

5.10   WATER QUALITY MONITORING IN SAMANA BLOCK

On the basis of observed values as given in Table No.5.8 and Table No.5.16, Samana block has pH value at 7.60 ranging from 7.14 to 7.89 with standard deviation 0.2262. Electrical Conductivities ranges from 0.68 to 1.97 with mean value 1.26 and standard deviation 0.3888 and Total dissolved Solids ranged from 435 to 1263 with mean value at 805, standard deviation 248.821. Total Suspended Solids ranges from 0.340 to 0.987 with mean at 0.629 and standard deviation 0.1944.

Total Alkalinity ranged from 169 to 469 with the mean at 305 and standard deviation 92.9541. Total Hardness, Calcium and Magnesium for this block are 403, 107.4 and 32.61 ranging from 218 to 632, 58 to 168 and 18 to 51 with standard deviations 124.411, 33.1762 and 10.0773 respectively.

Fluoride ranges from 0.176 to 0.512 with mean value of 0.326 and standard deviation 0.1007. The values of Chloride are 56.70 from 30 to 91, Nitrate at 2.023 from 1.093 to 3.173 with standard deviations 17.9593 and 0.6252 respectively.

Samana block has shown Total Coliform contents at 54 with a range from 29 to 85 and standard deviation 16.6715.

5.11   WATER QUALITY MONITORING IN SANOUR BLOCK

Water monitoring data regarding the rural drinking water supplies in this block is available in Table No.5.9 and Table No.5.17. pH is at 8.15 for this block ranging from 7.99 to 8.34 with standard deviation 0.1002. Electrical conductivities range from 0.75 to 2.08 with a mean 1.52 and standard deviation 0.3655 and Total Dissolved Solids show a value of 972 from 483 to 1329 with standard deviation at 233.914. Total Suspended Solids range from 0.377 to 1.038 with a mean value of 0.760 and standard deviation of 0.1827.

Total Alkalinity is at 396 for this block ranging from 197 to 543, standard deviation at 94.4830. Total Hardness ranges from 193 to 532 with a mean at 389 and standard deviation 93.5657. Calcium contents of this block are at 69.5 from 35 to 95, standard deviation 16.7082 and Magnesium is at 52.32 ranging from 26 to 72 with standard deviation of 12.5863.

Chlorides range from 42 to 117, mean value 85.32, standard deviation 20.7437 and Fluoride ranges from 0.130 to 0.358 with mean value of 0.262 and standard deviation 0.06293. Nitrates show variations from 1.621 to 4.460 with mean values of 3.263, standard deviation of 0.7850 and Total coliforms with mean value 100 range from 50 to 137 with standard deviation 24.1522.

Figures No.5.4 to 5.15 discusses the comparative study of different block on the basis of water quality parameters. Figures No. 5.16 to 5.27 discuss the comparative studies with respect to Minimum, Maximum, Mean and Standard Deviation of the water quality parameters.

GROUP AND CLUSTER-WISE STUDY

5.12   WATER QUALITY MONITORING IN GROUP - A

Group-A constitute the areas from which Ghaggar River passes. Ghaggar River carries waste water mixed with domestic, industrial and agricultural wastes. During the year, it does not carry much water but in the months of raining it comes with floods and disasters. Whatever it carries, it passes to the soil by leaching and deteriorate the quality of underground water. After passing through the Mohali District, it enters in the District Patiala in Ghanaur block, Bhunerheri block and then enters Haryana.

On the basis of data observed as given in Table No.5.20, Mean value for pH of this group is 8.05 with standard deviation 0.0707 indicating alkaline nature of water. Electrical Conductivity is at 1.40 with standard deviation 0.1414. Total Dissolved Solids for this block are at 920 with standard deviation 94.7523. Total Suspended Solids show the value of 0.578 with standard deviation 0.0028.

Total Alkalinity is 283 with standard deviation 50.9117. Total Hardness for this group is 269 with standard deviation 13.4350. Concentration of Calcium is 37.5 with standard deviation 12.3037 and Magnesium is at 42.4 with standard deviation 10.8187.

The water samples show the mean value for this group for Fluorine at 0.400 with standard deviation 0.1414. Chloride concentration is 84.55 with standard deviation 21.4253. Nitrate has mean value for this group as 2.300 with standard deviation 0.2828. Total Coliforms stand at 93 with standard deviation 9.1924.

5.13   WATER QUALITY MONITORING IN GROUP - B

Group-B constitutes the blocks Patiala and Sanour from where Patiale-wali Nadi passes. This Nadi also carries waste water but this Nadi behaves much like ponds with less water. Ghaggar carries continuous flow of water, although less but this Nadi does not have continuous flow of water. With the addition of domestic/ industrial/ agricultural wastes, this Nadi presents very deteriorating picture. In the rainy days, Patiale-wali Nadi becomes disastrous for the people of this area because of flood.

As per the observed data as in Table No.5.21, Mean pH stands at 8.0 for this group with standard deviation 0.2475. Electrical Conductivity value is 1.47 with standard deviation 0.0707. Total Dissolved Solids show a mean value of

942 with standard deviation 42.4264. The concentration of Total Suspended Solids was 0.736 with standard deviation 0.0339.

Total Alkalinity value is 373 with standard deviation 33.2340. Total Hardness show a value of 423 with standard deviation 47.3761, Calcium show a value of 95.5 with standard deviation 36.7695 and Magnesium stands at 44.61 with standard deviation 10.9036.

Fluoride concentration is found to be 0.316 with standard deviation 0.0757, Chloride is 112.46 with standard deviation 38.3818, Nitrate concentration is 3.160 with standard deviation 0.1457 and Total Coliforms are 91 with standard deviation 12.7279.

5.14   WATER QUALITY MONITORING IN CLUSTER - 1

Cluster-1 includes the areas come under the blocks Bhunerheri, Ghanaur, Patran, Patiala and Sanour where both Ghaggar River and Patiale-wali Nadi passes. After passing through Ghanaur and Bhunerheri, Ghaggar enters Haryana and after moving along the sides of Samana it enters Patran. Patiale-wali Nadi passes from Patiala, then Sanour and joins Ghaggar along the sides of Samana. In Patran both passes jointly as Ghaggar River and enters in Sangrur District. So in cluster-1 areas come under the blocks from where these rivers pass has been studied. As per the observed data as available in Table No.5.22 and Correlation Matrix available in Table No.5.23, the water quality parameters regarding this cluster may be discussed as given below:

5.14.1 pH

Mean pH for this cluster is found to be 8.02 with standard deviation 0.1351 indicating somewhat alkaline nature of water. The value obtained is within the desirable limit suggested by BIS (6.5 to 8.5) and WHO (7.0 to 8.5).

pH of this area is positively correlated with Electrical Conductivity (r = 0.0852), Total Dissolved Solids (r = 0.1039), Total Alkalinity (r = 0.2117), Magnesium (r = 0.7286), Nitrates (r = 0.0270), Total Coliforms (r = 0.3694) and negatively correlated with Total Suspended Solids (r = -0.0081), Total Hardness (r = -0.3345), Calcium (r = -0.5184), Fluoride (r = -0.4319), Chloride (r = -0.8637).

5.14.2 ELECTRICAL CONDUCTIVITY

Mean Electrical Conductivity for this cluster is found to be 1.49 with standard deviation 0.0912.

Electrical Conductivity of this area is positively correlated with pH (r = 0.0852), Total Dissolved Solids (r = 0.9771), Total Alkalinity (r = 0.2046), Total Suspended Solids (r = 0.5403), Total Hardness (r = 0.3412), Calcium (r = 0.3950), Fluoride (r = 0.3109), Chloride (r = 0.0419), Nitrate (r = 0.5737), Total Coliforms (r = 0.5943) and negatively correlated with Magnesium (r = -0.2856).

5.14.3 TOTAL DISSOLVED SOLIDS

Mean value of Total Dissolved Solids obtained in this cluster is 937 with standard deviation 54.8425 indicating more value than desirable limit set by BIS (500) and WHO (500). Although BIS suggests that water up to TDS value 2000 can be consumed if no alternative source is available but high TDS may change the taste of water and may also induce gastro-intestinal troubles.

In the present study, Total Dissolved Solids positively correlate with pH (r = 0.1039), Electrical Conductivity (r = 0.9771), Total Alkalinity (r = 0.0018), Total Suspended Solids (r = 0.3497), Total Hardness (r = 0.1564), Calcium (r = 0.2473), Fluoride (r = 0.4535), Chloride (r = 0.0232), Nitrate (r = 0.4516), total Coliforms (r = 0.5631) and negatively correlate with Magnesium (r = -0.3551).

5.14.4 TOTAL ALKALINITY

The mean value obtained for Total Alkalinity in this cluster is found to be 340 with standard deviation 60.2395 which is within the limits set by BIS (Desirable up to 200 but permissible up to 600). However, Total Alkalinity above 200 changes the taste of water to unpleasant and colour of the boiled rice to yellowish. It also hardens the rice and pulses after boiling.

In the present study, Total Alkalinity is positively correlated with pH (r = 0.2117), Electrical Conductivity (r = 0.2046), Total Dissolved Solids (r = 0.0018), Total Suspended Solids (r = 0.9053), Total Hardness (r = 0.7934), Calcium (r = 0.5849), Magnesium (r = 0.4662), Nitrate (r = 0.6610), Total Coliforms (r = 0.3600) and negatively correlated with Fluoride (r = -0.7121), Chloride (r = -0.2074).

5.14.5 TOTAL SUSPENDED SOLIDS

The mean value obtained for this cluster is 0.676 with standard deviation 0.0913. Although the Total Suspended Solids does not harm the human health but these are somewhat harmful to aquatic life as these stop the deep penetration of light in water and also affect the quality of water.

Total Suspended Solids are positively correlated with Electrical Conductivity (r = 0.5403), Total Dissolved Solids (r = 0.3497), Total Alkalinity (r = 0.9053), Total Hardness (r = 0.8960), Calcium (r = 0.7739), Magnesium (r = 0.1414), Chloride (r = 0.0367), Nitrate (r = 0.7821), Total Coliforms (r = 0.4472) and negatively correlated with pH (r = -0.0081), fluoride (r = -0.4137).

5.14.6 TOTAL HARDNESS

The mean value of Total Hardness obtained for this cluster is 373 with standard deviation 101.0312 which is within the limit suggested by BIS (Desirable up to 300 but permissible up to 600) and much above WHO (100). Besides difficulty in lathering of soaps, High Hardness of water is responsible for deteriorating the quality of clothes and skin irritation.

Total Hardness is positively correlated with Electrical Conductivity (r = 0.3412), Total Dissolved Solids (r = 0.1564), Total Alkalinity (r = 0.7934), Total Suspended Solids (r = 0.8960), Calcium (r = 0.9568), Chloride (r = 0.1684), Nitrate (r = 0.8255), Total Coliforms (r = 0.4189) and negatively correlated with pH (r = -0.3345), Magnesium (r = -0.1584), Fluoride (r = -0.2112).

5.14.7 CALCIUM

The mean value of Calcium in this cluster is found to be 78.9 with standard deviation 44.5049 which is within the limits suggested by BIS (Desirable up to 75 but permissible up to 200) and WHO (75). High Calcium contents leads to poor lathering of soaps and deteriorate the quality of cloths.

Calcium in this study is positively correlated with Electrical Conductivity (r = 0.3950), Total Dissolved Solids (r = 0.2473), Total Alkalinity (r = 0.5849), Total Suspended Solids (r = 0.7739), Total Hardness (r = 0.9568), Fluoride (r = 0.0744), Chloride (r = 0.2921), Nitrate (r = 0.8147), Total Coliforms (r = 0.4322) and negatively correlated with pH (r = -0.5184), Magnesium (r = -0.4387).

5.14.8 MAGNESIUM

The mean value obtained for Magnesium in this cluster is 42.58 with standard deviation 8.0216 which is within the limits prescribed by BIS (Desirable up to 30 but permissible up to 75) and WHO (50). High Magnesium contents leads to poor lathering of soaps and deteriorate the quality of cloths, with sulphate, Magnesium acts like laxative.

Magnesium in the present study is positively correlated with pH (r = 0.7286), Total Alkalinity (r = 0.4662), Total Suspended Solids (r = 0.1414), and negatively correlated with Electrical Conductivity (r = -0.2856), Total

Dissolved Solids (r = -0.3551), Total Hardness (r = -0.1583), Calcium (r = -0.4387), Fluoride (r = -0.9041), Chloride (r = -0.4745), Nitrate (r = -0.2131), and Total Coliforms (r = -0.1694).

5.14.9 FLUORIDE

The mean value obtained for Fluoride in this cluster is 0.364 with standard deviation 0.0918 which is within the limits suggested by BIS (Desirable up to 1.0 but permissible up to 1.5). High Fluoride leads to dental and skeletal fluorosis, so fluoride contents should remain as low as possible.

Fluoride in this study is positively correlated with Electrical Conductivity (r = 0.3109), Total Dissolved Solids (r = 0.4535), Calcium (r = 0.0744), Chloride (r = 0.2407), Nitrate (r = 0.0290) and Total Coliforms (r = 0.2098) and negatively correlated with pH (r = -0.4319), Total Alkalinity (r = -0.7121), Total Suspended Solids (r = -0.4137), Total Hardness (r = -0.2112), Magnesium (r = -0.9041).

5.14.10   CHLORIDE

In this study for this cluster the mean value for Chloride was found to be 91.56 with standard deviation 30.3168 which is within the limits prescribed by BIS (Desirable up to 250 but permissible up to 1000).

Chloride in this study is positively correlated with Electrical Conductivity (r = 0.0419), Total Dissolved Solids (r = 0.0232), Total Suspended Solids (r = 0.0367), Total Hardness (r = 0.1684), Calcium (r = 0.2921), Fluoride (r = 0.2407), and negatively correlated with pH (r = -0.8637), Total Alkalinity (r = -0.2074), Magnesium (r = -0.4745), Nitrate (r = -0.2707) and Total Coliforms (r = -0.6113).

5.14.11   NITRATE

The mean value obtained for Nitrate in this cluster is 3.156 with standard deviation 1.0568 which is within the prescribed limits by BIS (Desirable up to 45 but permissible up to 100).

Nitrate in the study of this cluster is positively correlated with pH (r = 0.0270), Electrical Conductivity (r = 0.5737), Total Dissolved Solids (r = 0.4516), Total Alkalinity (r = 0.6610), Total Suspended Solids (r = 0.7821), Total Hardness (r = 0.8255), Calcium (r = 0.8147), Fluoride (r = 0.0290), Total Coliforms (r = 0.8550) and negatively correlated with Magnesium (r = -0.2131), Chloride (r = -0.2707).

5.14.12   TOTAL COLIFORMS

The mean value obtained for Total Coliforms in the study of this cluster is 100 with standard deviation 18.8361. BIS suggests that drinking water should be free from pathogens to avoid water borne diseases like diarrhea, cholera, hepatitis etc. Water is classified into five types as per needs. Class-A, B and C type of water can be used for drinking purposes and these types may contain up to 50, 500, and 5000 Total Coliforms. But water is to be consumed after disinfection and treatment if contains pathogens to avoid any type of water borne diseases.

Total coliforms in the study of this cluster is positively correlated with pH (r = 0.3694), Electrical Conductivity (r = 0.5943), Total Dissolved Solids (r = 0.5631), Total Alkalinity (r = 0.3600), Total Suspended Solids (r = 0.4472), Total Hardness (r = 0.4189), Calcium (r = 0.4322), Fluoride (r = 0.2098), Nitrate (r = 0.8580), and negatively correlated with Magnesium (r = -0.1694), Chloride (r = -0.6113).

5.15   WATER QUALITY MONITORING IN CLUSTER - 2

Cluster-2 includes the areas come under the blocks Nabha, Rajpura and Samana which are away from both Ghaggar River and Patiale-wali Nadi. Sirhind canal passes through the Nabha block which merges in Bhakra canal passing through Patiala, Samana and Patran blocks before entering into Sangrur district. So in cluster-2 areas come under the blocks Nabha, Rajpura and Samana have been studied. On the basis of water monitoring data as observed and available in Table No.5.24 and Correlation Matrix as given in Table No.5.25, the water quality parameters may be discussed as given below:

5.15.1 pH

Mean pH for this cluster is found to be 7.78 with standard deviation 0.2720 indicating slightly alkaline nature of water. The value obtained is within the desirable limit suggested by BIS (6.5 to 8.5) and WHO (7.0 to 8.5).

pH of this area is positively correlated with Total Alkalinity (r = 0.0087), Magnesium (r = 0.9860), Fluoride (r = 0.7008), Chloride (r = 0.1391) and negatively correlated with Electrical Conductivity (r = -0.6666), Total Dissolved Solids (r = -0.6367), Total Suspended Solids (r = -0.6361), Total Hardness (r = -0.8291), Calcium (r = -0.9979), Nitrates (r = -0.4661), Total Coliforms (r = -0.5240).

5.15.2 ELECTRICAL CONDUCTIVITY

Mean Electrical Conductivity for this cluster is found to be 1.17 with standard deviation 0.0757.

Electrical Conductivity of this area is positively correlated with Total Dissolved Solids (r = 0.9992), Total Alkalinity (r = 0.7396), Total Suspended Solids (r = 0.9992), Total Hardness (r = 0.9695), Calcium (r = 0.7131), Fluoride (r = 0.0646), and negatively correlated with pH (r = -0.6666), Magnesium (r = -0.5332), Chloride (r = -0.8309), Nitrate (r = -0.3488), Total Coliforms (r = -0.2856).

5.15.3 TOTAL DISSOLVED SOLIDS

Mean value of Total Dissolved Solids obtained in this cluster is 749 with standard deviation 48.4183 indicating more value than desirable limit set by BIS (500) and WHO (500). Although BIS suggests that water up to TDS value 2000 can be consumed if no alternative source is available but high TDS may change the taste of water and may also induce gastro-intestinal troubles.

In the present study, Total Dissolved Solids positively correlate with Electrical Conductivity (r = 0.9992), Total Alkalinity (r = 0.7655), Total Suspended Solids (r = 1.0000), Total Hardness (r = 0.9590), Calcium (r = 0.6949), Fluoride (r = 0.1038), and negatively correlate with pH (r = -0.6367), Magnesium (r = -0.4995), Chloride (r = -0.8521), Nitrate (r = -0.3855), Total Coliforms (r = -0.3231).

5.15.4 TOTAL ALKALINITY

The mean value obtained for Total Alkalinity in this cluster is found to be 292 with standard deviation 14.0475 which is within the limits set by BIS (Desirable up to 200 but permissible up to 600). However, Total Alkalinity above 200 changes the taste of water to unpleasant and colour of the boiled rice to yellowish. It also hardens the rice and pulses after boiling.

In the present study, Total Alkalinity is positively correlated with pH (r = 0.0087), Electrical Conductivity (r = 0.7396), Total Dissolved Solids (r = 0.7655), Total Suspended Solids (r = 0.7660), Total Hardness (r = 0.5519), Calcium (r = 0.0555), Magnesium (r = 0.1750), Fluoride (r = 0.7194), and negatively correlated with Chloride (r = -0.9890), Nitrate (r = -0.8888), Total Coliforms (r = -0.8562).

5.15.5 TOTAL SUSPENDED SOLIDS

The mean value obtained for this cluster is 0.585 with standard deviation 0.0380. Although the Total Suspended Solids does not harm the human health but these are somewhat harmful to aquatic life as these stop the deep penetration of light in water and also affect the quality of water.

Total Suspended Solids are positively correlated with Electrical Conductivity (r = 0.9992), Total Dissolved Solids (r = 1.0000), Total Alkalinity (r = 0.7660), Total Hardness (r = 0.9588), Calcium (r = 0.6843), Fluoride (r = 0.1046) and negatively correlated with pH (r = -0.6361), Magnesium (r = - 0.4988), Chloride (r = -0.8526), Nitrate (r = -0.3862), Total Coliforms (r = - 0.3238).

5.15.6 TOTAL HARDNESS

The mean value of Total Hardness obtained for this cluster is 368 with standard deviation 32.3316 which is within the limit suggested by BIS (Desirable up to 300 but permissible up to 600) and much above WHO (100). Besides difficulty in lathering of soaps, High Hardness of water is responsible for deteriorating the quality of clothes and skin irritation.

Total Hardness is positively correlated with Electrical Conductivity (r = 0.9695), Total Dissolved Solids (r = 0.9590), Total Alkalinity (r = 0.5519), Total Suspended Solids (r = 0.9588), Calcium (r = 0.8632), and negatively correlated with pH (r = -0.8291), Magnesium (r = -0.7244), Fluoride (r = - 0.1821), Chloride (r = -0.6690), Nitrate (r = -0.1083), Total Coliforms (r = - 0.0418).

5.15.7 CALCIUM

The mean value of Calcium in this cluster is found to be 80.03 with standard deviation 38.5863 which is within the limits suggested by BIS (Desirable up to 75 but permissible up to 200) and WHO (75). High Calcium contents leads to poor lathering of soaps and deteriorate the quality of cloths.

Calcium in this study is positively correlated with Electrical Conductivity (r = 0.7131), Total Dissolved Solids (r = 0.6849), Total Alkalinity (r = 0.0555), Total Suspended Solids (r = 0.6843), Total Hardness (r = 0.8632), Nitrate (r = 0.4083), Total Coliforms (r = 0.4683) and negatively correlated with pH (r = -0.9979), Magnesium (r = -0.9733), Fluoride (r = -0.6536), Chloride (r = -0.2023).

5.15.8 MAGNESIUM

The mean value obtained for Magnesium in this cluster is 40.87 with standard deviation 17.1619 which is within the limits prescribed by BIS (Desirable up to 30 but permissible up to 75) and WHO (50). High Magnesium contents leads to poor lathering of soaps and deteriorate the quality of cloths and with sulphate, Magnesium acts like laxative.

Magnesium in the present study is positively correlated with pH (r = 0.9560), Total Alkalinity (r = 0.1750), Fluoride (r = 0.8098) and negatively correlated with Electrical Conductivity (r = -0.5332), Total Dissolved Solids (r = -0.4995), Total Suspended Solids (r = -0.4988), Total Hardness (r = -0.7244), Calcium (r = -0.9733), Chloride (r = -0.0277), Nitrate (r = -0.6068), and Total Coliforms (r = -0.6585).

5.15.9 FLUORIDE

The mean value obtained for Fluoride in this cluster is 0.295 with standard deviation 0.1356 which is within the limits suggested by BIS (Desirable up to 1.0 but permissible up to 1.5). High Fluoride leads to dental and skeletal fluorosis, so fluoride contents should remain as low as possible.

Fluoride in this study is positively correlated with pH (r = 0.7008), Electrical Conductivity (r = 0.0646), Total Dissolved Solids (r = 0.1038), Total Alkalinity (r = 0.7194), Total Suspended Solids (r = 0.1046), Magnesium (r = 0.8098), and negatively correlated with Total Hardness (r = -0.1821), Calcium (r = -0.6536), Chloride (r = -0.6090), Nitrate (r = -0.9578) and Total Coliforms (r = -0.9748).

5.15.10   CHLORIDE

In this study for this cluster the mean value for Chloride was found to be 84.83 with standard deviation 27.1590 which is within the limits prescribed by BIS (Desirable up to 250 but permissible up to 1000).

Chloride in this study is positively correlated with pH (r = 0.1391), Nitrate (r = 0.8113) and Total Coliforms (r = 0.7706) and negatively correlated with Electrical Conductivity (r = -0.8309), Total Dissolved Solids (r = -0.8521), Total Alkalinity (r = -0.9890), Total Suspended Solids (r = -0.8526), Total Hardness (r = -0.6690), Calcium (r = -0.2023), Magnesium (r = -0.0277), Fluoride (r = -0.6090).

5.15.11. NITRATE

The mean value obtained for Nitrate in this cluster is 3.744 with standard deviation 3.1742 which is within the prescribed limits by BIS (Desirable up to 45 but permissible up to 100).

Nitrate in the study of this cluster is positively correlated with Calcium (r = 0.4083), Chloride (r = 0.8113), Total Coliforms (r = 0.9978), and negatively correlated with pH (r = -0.4661), Electrical conductivity (r = -0.3488), Total dissolved Solids (r = -0.3855), Total Alkalinity (r = -0.8888), Total Suspended Solids (r = -0.3862), Total Hardness (r = -0.1083), Magnesium (r = -0.6068), Fluoride (r = -0.9576).

5.15.12   TOTAL COLIFORMS

The mean value obtained for Total Coliforms in the study of this cluster is 96 with standard deviation 88.7919. BIS suggests that drinking water should be free from pathogens to avoid water borne diseases like diarrhea, cholera, hepatitis etc. Water is classified into five types as per needs. Class-A, B and C type of water can be used for drinking purposes and these types may contain up to 50, 500, and 5000 Total Coliforms. But water is to be consumed after disinfection and treatment if contains pathogens to avoid any type of water borne diseases.

Total Coliforms in the study of this cluster is positively correlated with Calcium (r = 0.4683), Chloride (r = 0.7706), Nitrate (r = 0.9978), and negatively correlated with pH (r = -0.5240), Electrical Conductivity (r = -0.2856), Total dissolved Solids (r = -0.3231), Total Alkalinity (r = -0.8562), Total Suspended Solids (r = -0.3238), Total Hardness (r = -0.0418), Magnesium (r = -0.6585), Fluoride (r = -0.9748).

Figure No.5.28 and Figure No.5.29 discuss the comparative study of water quality parameters of these two clusters.

5.16   WATER QUALITY MONITORING IN
DISTRICT PATIALA AS A WHOLE

District Patiala includes a total of eight blocks viz. Bhunerheri, Ghanaur, Patran, Patiala, Sanour, Nabha, Rajpura and Samana. Ghaggar River, Patiale-wali Nadi, Sirhind canal and Bhakra canal passes through the district. Industrial development is progressive. Maximum area of the district is used for agriculture. Water development is very high as compared to the state. In this part of the study, the whole district, on the basis of the observations and analysis made as given in Table No.5.26 and Correlation Matrix Table No.5.27, is studied:

5.16.1 pH

Mean pH for the district is found to be 7.93 with standard deviation 0.2178 indicating slightly alkaline nature of water. The value obtained is within the desirable limit suggested by BIS (6.5 to 8.5) and WHO (7.0 to 8.5).

pH of this area is positively correlated with Electrical Conductivity (r = 0.4104), Total Dissolved Solids (r = 0.4341), Total Alkalinity (r = 0.3608), Total Suspended Solids (r = 0.2124), Magnesium (r = 0.7816), Fluoride (r = 0.3802) and negatively correlated with Total Hardness (r = -0.2587), Calcium (r = -0.5643), Chloride (r = -0.2164), Nitrates (r = -0.3653), Total Coliforms (r = -0.2647).

5.16.2 ELECTRICAL CONDUCTIVITY

Mean Electrical Conductivity for this cluster is found to be 1.345 with standard deviation 0.1631.

Electrical Conductivity of this area is positively correlated with pH (r = 0.4104), Total Dissolved Solids (r = 0.9951), Total Alkalinity (r = 0.5141), Total Suspended Solids (r = 0.7183), Total Hardness (r = 0.2183), Calcium (r = 0.2214), Fluoride (r = 0.3883), Chloride (r = 0.0163), Total Coliforms (r = 0.0352) and negatively correlated with Magnesium (r = -0.1066), Nitrate (r = -0.1150).

5.16.3 TOTAL DISSOLVED SOLIDS

Mean value of Total Dissolved Solids obtained in this cluster is 867 with standard deviation 108.8219 indicating more value than desirable limit set by BIS (500) and WHO (500). Although BIS suggests that water up to TDS value 2000 can be consumed if no alternative source is available but high TDS may change the taste of water and may also induce gastro-intestinal troubles.

In the present study, Total Dissolved Solids positively correlate with pH (r = 0.4331), Electrical Conductivity (r = 0.9951), Total Alkalinity (r = 0.4497), Total Suspended Solids (r = 0.6514), Total Hardness (r = 0.1336), Calcium (r = 0.1517), Fluoride (r = 0.4284), Chloride (r = 0.0135), Total Coliforms (r = 0.0198) and negatively correlate with Magnesium (r = -0.1016), Nitrate (r = -0.1527).

5.16.4 TOTAL ALKALINITY

The mean value obtained for Total Alkalinity in this cluster is found to be 322 with standard deviation 52.3967 which is within the limits set by BIS (Desirable up to 200 but permissible up to 600). However, Total Alkalinity above 200 changes the taste of water to unpleasant and colour of the boiled rice to yellowish. It also hardens the rice and pulses after boiling.

In the present study, Total Alkalinity is positively correlated with pH (r = 0.3608), Electrical Conductivity (r = 0.5141), Total Dissolved Solids (r = 0.4497), Total Suspended Solids (r = 0.9168), Total Hardness (r = 0.7031), Calcium (r = 0.4302), Magnesium (r = 0.2815), Nitrate (r = 0.0520) and negatively correlated with Fluoride (r = -0.1746), Chloride (r = -0.1660), Total Coliforms (r = -0.0108).

5.16.5 TOTAL SUSPENDED SOLIDS

The mean value obtained for this cluster is 0.642 with standard deviation 0.0859. Although the Total Suspended Solids does not harm the human health but these are somewhat harmful to aquatic life as these stop the deep penetration of light in water and also affect the quality of water.

Total Suspended Solids are positively correlated with pH (r = 0.2124), Electrical Conductivity (r = 0.7183), Total Dissolved Solids (r = 0.6514), Total Alkalinity (r = 0.9168), Total Hardness (r = 0.7673), Calcium (r = 0.6062), Magnesium (r = 0.0081), Nitrate (r = 0.0952), Total Coliforms (r = 0.0493) and negatively correlated with Fluoride (r = -0.0169), Chloride (r = -0.0127).

5.16.6 TOTAL HARDNESS

The mean value of Total Hardness obtained for this cluster is 371 with standard deviation 78.3345 which is within the limit suggested by BIS (Desirable up to 300 but permissible up to 600) and much above WHO (100). Besides difficulty in lathering of soaps, High Hardness of water is responsible for deteriorating the quality of clothes and skin irritation.

Total Hardness is positively correlated with Electrical Conductivity (r = 0.2183), Total Dissolved Solids (r = 0.1336), Total Alkalinity (r = 0.7031), Total Suspended Solids (r = 0.7673), Calcium (r = 0.8942), Chloride (r = 0.0628), Nitrate (r = 0.3125), Total Coliforms (r = 0.1095) and negatively correlated with pH (r = -0.2587), Magnesium (r = -0.2155, Fluoride (r = -0.1521).

5.16.7 CALCIUM

The mean value of Calcium in this cluster is found to be 79.3 with standard deviation 39.4664 which is within the limits suggested by BIS (Desirable up to 75 but permissible up to 200) and WHO (75). High Calcium contents leads to poor lathering of soaps and deteriorate the quality of cloths.

Calcium in this study is positively correlated with Electrical Conductivity (r = 0.2214), Total Dissolved Solids (r = 0.1517), Total Alkalinity (r = 0.4302), Total Suspended Solids (r = 0.6062), Total Hardness (r= 0.8942), Chloride (r = 0.1505), Nitrate (r = 0.4850), Total Coliforms (r = 0.3395) and negatively correlated with pH (r = -0.5643), Magnesium (r = -0.6298), Fluoride (r = -0.1964).

5.16.8 MAGNESIUM

The mean value obtained for Magnesium in this cluster is 41.94 with standard deviation 11.0318 which is within the limits prescribed by BIS (Desirable up to 30 but permissible up to 75) and WHO (50). High Magnesium contents leads to poor lathering of soaps and deteriorate the quality of cloths and with sulphate, Magnesium acts like laxative.

Magnesium in the present study is positively correlated with pH (r = 0.7816), Total Alkalinity (r = 0.2815), Total Suspended Solids (r = 0.0081), Fluoride (r = 0.1614) and negatively correlated with Electrical Conductivity (r = -0.1066), Total Dissolved Solids (r = -0.1016), Total Hardness (r= -0.2155), Calcium (r = -0.6298), Chloride (r = -0.2206), Nitrate (r = -0.5127), and Total Coliforms (r = -0.5480).

5.16.9 FLUORIDE

The mean value obtained for Fluoride in this cluster is 0.338 with standard deviation 0.1065 which is within the limits suggested by BIS (Desirable up to 1.0 but permissible up to 1.5). High Fluoride leads to dental and skeletal fluorosis, so fluoride contents should remain as low as possible.

Fluoride in this study is positively correlated with pH (r = 0.3802), Electrical Conductivity (r = 0.3883), Total Dissolved Solids (r = 0.4284), Magnesium (r = 0.1614), and negatively correlated with Total Alkalinity (r = -0.1746), Total Suspended Solids (r = -0.0169), Total Hardness (r= -0.1521), Calcium (r = -0.1964), Chloride (r = -0.0460), Nitrate (r = -0.6280) and Total Coliforms (r = -0.5841).

5.16.10   CHLORIDE

In this study for this cluster the mean value for Chloride was found to be 89.04 with standard deviation 27.3507 which is within the limits prescribed by BIS (Desirable up to 250 but permissible up to 1000).

Chloride in this study is positively correlated with Electrical Conductivity (r = 0.0163), Total Dissolved Solids (r = 0.0135), Total Hardness (r= 0.0628), Calcium (r = 0.1505), Nitrate (r = 0.2688) and Total Coliforms (r = 0.2489), and negatively correlated with pH (r = -0.2164), Total Alkalinity (r = -0.1660), Total Suspended Solids (r = -0.0127), Magnesium (r = -0.2206), Fluoride (r = -0.0460).

5.16.11   NITRATE

The mean value obtained for Nitrate in this cluster is 3.376 with standard deviation 1.8999 which is within the prescribed limits by BIS (Desirable up to 45 but permissible up to 100).

Nitrate in the study of this cluster is positively correlated with Total Alkalinity (r = 0.0520), Total Suspended Solids (r = 0.0952), Total Hardness (r= 0.3125), Calcium (r = 0.4850), Chloride (r = 0.2688), Total Coliforms (r = 0.9509) and negatively correlated with pH (r = -0.3653), Electrical Conductivity (r = -0.1150), Total Dissolved Solids (r = -0.1527), Magnesium (r = -0.5127), Fluoride (r = -0.6280).

5.16.12   TOTAL COLIFORMS

The mean value obtained for Total Coliforms in the study of this cluster is 98 with standard deviation 49.5024. BIS suggests that drinking water should be free from pathogens to avoid water borne diseases like diarrhea, cholera, hepatitis etc. Water is classified into five types as per needs. Class-A, B and C type of water can be used for drinking purposes and these types may contain up to 50, 500, and 5000 Total Coliforms. But water is to be consumed after disinfection and treatment if contains pathogens to avoid any type of water borne diseases.

Total Coliforms in the study of this cluster is positively correlated with Electrical Conductivity (r = 0.0352), Total Dissolved Solids (r = 0.0198), Total Suspended Solids (r = 0.0493), Total Hardness (r= 0.1095), Calcium (r = 0.3395), Chloride (r = 0.2489), Nitrate (r = 0.9509) and negatively correlated with pH (r = -0.2647), Total Alkalinity (r = -0.0108), Magnesium (r = -0.5480), Fluoride (r = -0.5841).

5.17 WATER QUALITY PARAMETERS OF GENERAL IMPORTANCE

Two quality parameters, TOTAL DISSOLVED SOLIDS (Table No.5.19 and Figure No.5.30 and TOTAL HARDNESS (Table No.5.18 and Figure No.5.2) are of general importance. On the basis of physicochemical monitoring of the samples studied, it may be concluded that the water of all the samples studied is of hard type and none of the samples showed soft or moderately hard type water quality. Water with Total Hardness below 75 can be considered soft, 75 to 150 as moderately hard, 151 to 300 as hard and water with Total Hardness above 300 is termed as very hard (Matalas and Reiher, 1967 & Sawyer and McCarthy, 1967). Table No. 5.18 and Figure No. 5.2 indicates that 80% is hard water and 20% of water is very hard in the Bhunerheri block. Ghanaur block has 67% hard water and 33% very hard water. Water of Nabha block can be classified as 50% hard and 50% very hard. Rajpura block has 44% hard and 56% very hard water. Patiala block has 25% hard and 75% very hard water. Patran block contains 25% hard and 75% very hard water. Water of Samana block can be classified as 20% hard and 80% very hard. Sanour block contains 10% hard water and 90% very hard water. Consumption of very soft and very hard water is not good for human beings. Very soft water does not contain much solids dissolved, needed to digest the food while very hard water contains much solids dissolved in it and can create problems. So the water samples come under Slightly Hard and Moderately Hard category, can be consumed safely, if otherwise found suitable. As per Total Dissolved Solids are concerned, safe and desirable limit as prescribed by BIS and WHO is 500, but water with Total Dissolved Solids up to 1500 (WHO) and 2000 (BIS) can be consumed if no other alternative is available. On the basis of physicochemical monitoring of the samples studied (Table No.5.19 and Figure No.5.3), it may be concluded that none of the samples showed Total Dissolved Solids below 300 means no water can be termed as good. Water with Total Dissolved Solids 300 to 500 can be considered fair, 500 to 900 average, 900 to 1200 poor, 1200 to 2000 very poor and water with Total Dissolved Solids above 2000 is termed as unacceptable (Matalas and Reiher, 1967). On the basis of Total Dissolved Solids, it was found that none water samples in Bhunerheri block, 17% water samples in Ghanaur block, 44% water samples in Nabha block, 15% water samples in Patran block, 25% water samples in Patiala block, 33% water samples in Rajpura block, 10% water samples in Samana block and 5% water samples in Sanour block may be termed as fair. Average plus poor category include 80% water samples in Bhunerheri block, 70% water samples in Ghanaur block, 44% water samples in Nabha block, 60% water samples in Patran block, 60% water samples in Patiala block, 57% water samples in Rajpura block, 85% water samples in Samana block and 75% water samples in Sanour block. All the block include some water samples in very poor category and one water samples was found in Ghanaur block in unacceptable category.

Sample	T °C	pH	EC dS/m	TDS mg/L	TA mg/L	TSS mg/L	TH mg/L	Ca^{2+} mg/L	Mg^{2+} mg/L	F^- mg/L	Cl^- mg/L	NO_3^- mg/L	Coliforms MPN/100ml
Bh-01	22.3	7.93	1.37	878	220	0.5719	248	42	34	0.474	109	2.206	88
Bh-02	22.2	7.98	1.35	867	217	0.5745	214	49	23	0.207	103	2.178	87
Bh-03	22.1	8.02	1.96	1253	313	0.5766	312	48	47	0.687	97	3.148	126
Bh-04	22.0	8.04	1.48	947	237	0.5777	285	40	45	0.699	94	2.379	95
Bh-05	22.8	7.78	2.44	1563	391	0.5639	289	65	31	0.287	116	3.927	157
Bh-06	22.4	7.91	1.33	849	212	0.5708	213	48	23	0.208	109	2.133	85
Bh-07	21.8	8.13	1.33	853	213	0.5825	336	30	63	1.030	87	2.143	86
Bh-08	22.0	8.03	1.39	887	222	0.5772	216	49	22	0.199	97	2.229	89
Bh-09	22.2	7.96	1.39	889	222	0.5735	267	40	41	0.610	102	2.234	89
Bh-10	22.9	7.72	2.13	1361	340	0.5607	265	62	27	0.227	117	3.420	137
Bh-11	22.1	8.02	1.36	868	217	0.5766	215	48	23	0.215	98	2.181	87
Bh-12	22.2	7.98	1.34	859	215	0.5745	214	48	23	0.214	103	2.158	86
Bh-13	22.0	8.06	1.50	962	241	0.5788	238	49	28	0.322	93	2.417	97
Bh-14	22.2	7.98	1.36	869	217	0.5745	214	49	23	0.205	103	2.183	87
Bh-15	21.8	8.12	1.35	863	216	0.5820	293	35	50	0.816	86	2.168	87
Bh-16	21.7	8.14	1.29	825	206	0.5830	313	32	57	0.977	97	2.073	83
Bh-17	21.6	8.21	2.28	1456	364	0.5868	335	52	50	0.730	96	3.658	146
Bh-18	21.8	8.12	1.86	1189	297	0.5820	293	49	42	0.586	97	2.987	119
Bh-19	21.6	8.18	1.50	963	241	0.5852	321	36	56	0.940	96	2.420	97
Bh-20	21.5	8.23	1.30	829	207	0.5878	221	45	26	0.300	95	2.083	83
Bh-21	21.4	8.26	2.01	1286	322	0.5894	331	47	52	0.806	103	3.231	129
Bh-22	22.1	8.02	1.28	818	205	0.5766	215	46	25	0.263	98	2.055	82
Bh-23	21.5	8.24	1.30	830	208	0.5884	221	45	26	0.302	95	2.085	83
Bh-24	22.3	7.94	1.50	963	241	0.5724	283	41	44	0.673	99	2.420	97
Bh-25	22.2	7.98	1.83	1173	293	0.5745	256	55	29	0.298	98	2.947	118
Bh-26	23.0	7.69	1.25	801	200	0.5591	207	47	22	0.203	102	2.013	81
Bh-27	22.5	7.87	1.97	1263	316	0.5687	289	52	38	0.502	100	3.173	127
Bh-28	22.4	7.89	1.27	812	203	0.5697	212	46	24	0.239	103	2.040	82
Bh-29	22.2	7.98	1.28	816	204	0.5745	214	46	24	0.256	98	2.050	82
Bh-30	22.3	7.93	1.27	814	204	0.5719	213	46	24	0.246	99	2.045	82
Mean	**22.1**	**8.00**	**1.50**	**987**	**247**	**0.5762**	**259**	**46.2**	**34.7**	**0.500**	**99.7**	**2.500**	**99**

Table 5.2: Physicochemical and microbiological characteristics of drinking water in *Bhunerheri* block of district Patiala

Sample	T °C	pH	EC dS/m	TDS mg/L	TA mg/L	TSS mg/L	TH mg/L	Ca^{2+} mg/L	Mg^{2+} mg/L	F^- mg/L	Cl^- mg/L	NO_3^- mg/L	Coliforms MPN/100ml
Gh-01	21.7	8.14	1.20	765	286	0.520	260	26	47	0.346	63	1.922	77
Gh-02	21.6	8.21	0.88	565	285	0.384	192	19	35	0.256	47	1.420	57
Gh-03	21.8	8.12	0.83	529	407	0.359	180	18	33	0.240	44	1.329	53
Gh-04	21.6	8.18	1.50	963	310	0.654	327	33	60	0.436	80	2.420	97
Gh-05	21.5	8.23	1.82	1167	515	0.793	396	40	72	0.529	98	2.932	117
Gh-06	21.4	8.26	0.90	576	281	0.391	196	20	36	0.261	48	1.447	58
Gh-07	22.1	8.02	1.37	874	274	0.594	297	30	54	0.396	71	2.196	88
Gh-08	21.5	8.24	1.18	756	292	0.514	257	26	47	0.342	63	1.899	76
Gh-09	22.3	7.94	1.05	673	282	0.457	229	23	42	0.305	54	1.691	68
Gh-10	22.2	7.98	2.15	1376	434	0.935	467	47	85	0.379	112	3.457	138
Gh-11	23.0	7.69	1.40	893	267	0.607	303	30	55	0.404	70	2.244	90
Gh-12	22.5	7.87	1.05	675	270	0.459	229	23	42	0.306	54	1.696	68
Gh-13	22.4	7.89	0.90	578	304	0.393	196	20	36	0.262	46	1.452	58
Gh-14	22.2	7.98	1.60	1023	277	0.695	347	35	63	0.463	83	2.570	103
Gh-15	22.3	7.93	1.53	982	274	0.667	334	33	61	0.445	79	2.467	99
Gh-16	22.2	7.96	2.45	1567	263	1.065	478	53	84	0.478	127	3.937	157
Gh-17	22.8	7.75	2.14	1367	451	0.929	464	46	85	0.563	108	3.435	137
Gh-18	22.5	7.87	3.70	2367	374	1.608	504	74	78	0.376	175	5.947	238
Gh-19	22.4	7.89	1.52	974	304	0.662	331	33	60	0.441	78	2.447	98
Gh-20	22.3	7.93	1.30	834	263	0.567	283	28	52	0.378	67	2.095	84
Gh-21	22.2	7.98	1.19	763	410	0.518	259	26	47	0.346	62	1.917	77
Gh-22	22.3	7.94	1.19	763	260	0.518	259	26	47	0.346	62	1.917	77
Gh-23	21.6	8.21	1.14	729	273	0.495	248	25	45	0.330	61	1.832	73
Gh-24	21.0	8.43	1.01	649	325	0.441	220	22	40	0.294	56	1.631	65
Gh-25	20.7	8.54	0.73	467	401	0.317	159	16	29	0.212	41	1.173	47
Gh-26	21.0	8.42	1.30	829	270	0.563	282	28	51	0.375	71	2.083	83
Gh-27	21.5	8.23	0.74	476	416	0.323	162	16	29	0.216	40	1.196	48
Gh-28	21.7	8.14	0.74	472	264	0.321	160	16	29	0.214	39	1.186	47
Gh-29	21.8	8.12	0.75	482	265	0.327	164	16	30	0.218	40	1.211	48
Gh-30	21.9	8.08	0.73	468	263	0.318	159	16	29	0.212	39	1.176	47
Mean	**21.9**	**8.10**	**1.30**	**853**	**319**	**0.580**	**278**	**28.8**	**50**	**0.300**	**69.4**	**2.100**	**86**

Table 5.3: Physicochemical and microbiological characteristics of drinking water in *Ghanaur* block of district Patiala

Sample	T °C	pH	EC dS/m	TDS mg/L	TA mg/L	TSS mg/L	TH mg/L	Ca^{2+} mg/L	Mg^{2+} mg/L	F^- mg/L	Cl^- mg/L	NO_3^- mg/L	Coliforms MPN/100ml
Nb-01	22.2	7.96	0.72	463	223	0.362	232	23	42	0.309	55	1.163	23
Nb-02	22.8	7.75	1.31	839	317	0.655	420	42	76	0.559	98	2.108	42
Nb-03	22.5	7.87	0.71	456	220	0.356	228	23	42	0.304	54	1.146	23
Nb-04	22.4	7.89	1.07	683	280	0.534	342	34	62	0.455	81	1.716	34
Nb-05	22.3	7.93	1.47	938	348	0.733	469	47	85	0.325	112	2.357	47
Nb-06	22.2	7.98	1.92	1227	426	0.959	514	61	88	0.418	147	3.083	62
Nb-07	22.3	7.94	2.15	1376	464	1.075	498	69	79	0.417	164	3.457	69
Nb-08	21.6	8.21	2.24	1431	492	1.118	516	72	82	0.454	176	3.595	72
Nb-09	21.0	8.43	1.31	837	335	0.654	419	42	76	0.558	106	2.103	42
Nb-10	20.7	8.54	0.73	467	233	0.365	234	23	43	0.311	60	1.173	23
Nb-11	21.0	8.42	0.72	462	230	0.361	231	23	42	0.308	58	1.161	23
Nb-12	21.5	8.23	0.74	476	231	0.372	238	24	43	0.317	59	1.196	24
Nb-13	21.7	8.14	0.83	534	245	0.417	267	27	49	0.356	65	1.342	27
Nb-14	21.8	8.12	0.75	482	230	0.377	241	24	44	0.321	59	1.211	24
Nb-15	21.9	8.08	1.53	982	364	0.767	491	49	89	0.655	119	2.467	49
Nb-16	21.5	8.23	1.15	736	302	0.575	368	37	67	0.491	91	1.849	37
Nb-17	21.2	8.34	1.45	928	358	0.725	464	46	85	0.619	116	2.332	47
Nb-18	21.7	8.14	1.31	836	327	0.653	418	42	76	0.557	102	2.101	42
Nb-19	21.6	8.18	0.71	453	224	0.354	227	23	41	0.302	56	1.138	23
Nb-20	21.8	8.12	0.99	635	272	0.496	318	32	58	0.423	77	1.595	32
Nb-21	21.9	8.08	0.84	536	244	0.419	268	27	49	0.357	65	1.347	27
Nb-22	21.6	8.19	0.74	473	229	0.370	237	24	43	0.315	58	1.188	24
Nb-23	21.9	8.09	0.69	439	218	0.343	220	22	40	0.293	53	1.103	22
Nb-24	22.0	8.04	1.17	746	300	0.583	373	37	68	0.497	90	1.874	37
Nb-25	22.1	8.02	0.71	457	222	0.357	229	23	42	0.305	55	1.148	23
Nb-26	22.2	7.98	0.78	498	232	0.389	249	25	45	0.332	60	1.251	25
Nb-27	22.2	7.96	0.76	487	229	0.380	244	24	44	0.325	58	1.224	24
Nb-28	22.3	7.93	1.48	948	351	0.741	474	47	86	0.632	113	2.382	48
Nb-29	22.4	7.92	1.92	1231	425	0.962	516	62	88	0.563	146	3.093	62
Nb-30	22.5	7.86	0.72	459	220	0.359	230	23	42	0.306	54	1.153	23
Mean	**21.9**	**8.09**	**1.12**	**717**	**293**	**0.560**	**339**	**35.9**	**60.6**	**0.413**	**86.9**	**1.802**	**36**

Table 5.4: Physicochemical and microbiological characteristics of drinking water in *Nabha* block of district Patiala

Sample	T °C	pH	EC dS/m	TDS mg/L	TA mg/L	TSS mg/L	TH mg/L	Ca^{2+} mg/L	Mg^{2+} mg/L	F^- mg/L	Cl^- mg/L	NO_3^- mg/L	Coliforms MPN/100ml
Rj-01	23.3	6.98	0.88	562	196	0.439	281	75	23	0.114	79	5.739	153
Rj-02	23.1	7.02	0.73	468	164	0.366	234	62	19	0.095	66	4.776	127
Rj-03	23.1	7.04	0.62	395	139	0.309	198	53	16	0.080	56	4.031	107
Rj-04	22.9	7.08	1.31	836	296	0.653	418	111	34	0.169	118	8.531	227
Rj-05	22.8	7.12	1.17	748	266	0.584	374	100	30	0.151	107	7.633	204
Rj-06	22.8	7.13	0.75	482	172	0.377	241	64	20	0.098	69	4.918	131
Rj-07	20.7	7.85	2.31	1476	579	1.153	738	197	60	0.299	232	15.061	402
Rj-08	21.5	7.54	1.01	647	244	0.505	324	86	26	0.131	98	6.602	176
Rj-09	22.3	7.28	1.32	846	308	0.661	423	113	34	0.171	123	8.633	230
Rj-10	22.2	7.3	0.91	584	213	0.456	292	78	24	0.118	85	5.959	159
Rj-11	22.2	7.31	2.23	1426	521	1.114	713	190	58	0.289	208	14.551	388
Rj-12	21.9	7.43	1.45	930	345	0.727	465	124	38	0.188	138	9.490	253
Rj-13	22.1	7.36	0.76	486	179	0.380	243	65	20	0.098	72	4.959	132
Rj-14	22.2	7.31	1.46	937	342	0.732	469	125	38	0.190	137	9.561	255
Rj-15	22.3	7.29	1.37	874	319	0.683	437	117	35	0.177	127	8.918	238
Rj-16	20.4	7.98	2.14	1367	545	1.068	684	182	55	0.277	218	13.949	372
Rj-17	20.4	7.98	1.47	938	374	0.733	469	125	38	0.190	150	9.571	255
Rj-18	20.2	8.03	1.47	938	377	0.733	469	125	38	0.190	151	9.571	255
Rj-19	20.0	8.1	0.75	478	194	0.373	239	64	19	0.097	77	4.878	130
Rj-20	20.0	8.14	0.73	467	190	0.365	234	62	19	0.095	76	4.765	127
Rj-21	19.9	8.17	0.59	376	154	0.294	188	50	15	0.076	61	3.837	102
Rj-22	20.7	7.84	1.15	739	290	0.577	370	99	30	0.150	116	7.541	201
Rj-23	20.1	8.09	0.91	583	236	0.455	292	78	24	0.118	94	5.949	159
Rj-24	20.2	8.03	1.14	729	293	0.570	365	97	30	0.148	117	7.439	198
Rj-25	21.1	7.68	0.76	485	186	0.379	243	65	20	0.098	74	4.949	132
Rj-26	20.2	8.03	0.73	467	188	0.365	234	62	19	0.095	75	4.765	127
Rj-27	20.2	8.04	1.03	658	265	0.514	329	88	27	0.133	106	6.714	179
Rj-28	19.9	8.15	1.14	729	297	0.570	365	97	30	0.148	119	7.439	198
Rj-29	20.7	7.83	1.01	648	254	0.506	324	86	26	0.131	101	6.612	176
Rj-30	20.0	8.12	0.75	478	194	0.373	239	64	19	0.097	78	4.878	130
Mean	**21.3**	**7.64**	**1.14**	**726**	**277**	**0.567**	**363**	**96.8**	**29.4**	**0.147**	**110.9**	**7.407**	**198**

Table 5.5: Physicochemical and microbiological characteristics of drinking water in *Rajpura* block of district Patiala

Sample	T °C	pH	EC dS/m	TDS mg/L	TA mg/L	TSS mg/L	TH mg/L	Ca^{2+} mg/L	Mg^{2+} mg/L	F^- mg/L	Cl^- mg/L	NO_3^- mg/L	Coliforms MPN/100ml
Pt-01	21.5	8.23	0.76	485	200	0.379	243	65	20	0.196	80	1.628	43
Pt-02	23.4	7.58	1.71	1092	414	0.853	546	146	44	0.442	166	3.664	98
Pt-03	23.7	7.47	1.79	1145	428	0.895	573	153	46	0.464	171	3.842	102
Pt-04	22.2	7.98	1.33	849	339	0.663	425	113	34	0.344	136	2.849	76
Pt-05	25.4	6.97	2.07	1326	462	1.036	663	177	54	0.537	185	4.450	119
Pt-06	21.9	8.08	1.69	1083	438	0.846	542	144	44	0.439	175	3.634	97
Pt-07	25.5	6.95	2.57	1643	571	1.284	822	219	67	0.665	228	5.513	147
Pt-08	21.0	8.42	0.74	473	199	0.370	237	63	19	0.192	80	1.587	42
Pt-09	24.7	7.16	2.00	1283	459	1.002	642	171	52	0.520	184	4.305	115
Pt-10	24.7	7.18	2.14	1367	491	1.068	684	182	55	0.554	196	4.587	122
Pt-11	22.2	7.98	1.48	948	378	0.741	474	126	38	0.384	151	3.181	85
Pt-12	22.2	7.96	1.01	645	257	0.504	323	86	26	0.261	103	2.164	58
Pt-13	22.3	7.93	1.31	839	333	0.655	420	112	34	0.340	133	2.815	75
Pt-14	22.4	7.92	1.17	749	297	0.585	375	100	30	0.303	119	2.513	67
Pt-15	22.0	8.04	1.45	930	374	0.727	465	124	38	0.377	150	3.121	83
Pt-16	22.5	7.87	1.82	1163	458	0.909	582	155	47	0.471	183	3.903	104
Pt-17	21.2	8.36	0.76	485	203	0.379	243	65	20	0.196	81	1.628	43
Pt-18	22.4	7.91	0.77	493	195	0.385	247	66	20	0.200	78	1.654	44
Pt-19	22.2	7.98	0.91	584	233	0.456	292	78	24	0.237	93	1.960	52
Pt-20	22.3	7.95	1.00	639	254	0.499	320	85	26	0.259	102	2.144	57
Mean	22.8	7.80	1.42	911	349	0.712	456	121.5	36.90	0.369	139.6	3.057	82

Table 5.6: Physicochemical and microbiological characteristics of drinking water in *Patiala* block of district Patiala

Sample	T °C	pH	EC dS/m	TDS mg/L	TA mg/L	TSS mg/L	TH mg/L	Ca^{2+} mg/L	Mg^{2+} mg/L	F^- mg/L	Cl^- mg/L	NO_3^- mg/L	Coliforms MPN/100ml
Pr-01	22.2	7.98	1.51	965	385	0.754	483	129	39	0.391	64	4.874	130
Pr-02	22.2	7.96	1.08	693	276	0.541	347	92	28	0.281	46	3.500	93
Pr-03	22.3	7.93	2.15	1374	545	1.073	687	183	56	0.556	92	6.939	185
Pr-04	22.4	7.92	2.46	1572	623	1.228	786	210	64	0.637	106	7.939	212
Pr-05	22.5	7.86	2.29	1467	577	1.146	734	196	59	0.594	100	7.409	198
Pr-06	22.5	7.87	1.46	937	369	0.732	469	125	38	0.379	63	4.732	126
Pr-07	22.2	7.96	1.00	638	254	0.498	319	85	26	0.258	43	3.222	86
Pr-08	22.4	7.91	0.93	593	235	0.463	297	79	24	0.240	40	2.995	80
Pr-09	22.2	7.98	0.75	482	192	0.377	241	64	20	0.195	32	2.434	65
Pr-10	22.3	7.95	0.75	483	192	0.377	242	64	20	0.196	32	2.439	65
Pr-11	22.2	7.98	1.68	1073	428	0.838	537	143	43	0.435	72	5.419	145
Pr-12	22.1	8.02	1.46	937	376	0.732	469	125	38	0.379	62	4.732	126
Pr-13	22.0	8.04	1.76	1128	453	0.881	564	150	46	0.457	75	5.697	152
Pr-14	21.9	8.07	1.92	1227	495	0.959	614	164	50	0.497	81	6.197	165
Pr-15	21.8	8.12	1.50	959	389	0.749	480	128	39	0.388	63	4.843	129
Pr-16	21.5	8.23	2.29	1467	604	1.146	734	196	59	0.594	95	7.409	198
Pr-17	21.2	8.34	1.71	1093	456	0.854	547	146	44	0.443	70	5.520	147
Pr-18	21.2	8.36	1.69	1084	453	0.847	542	145	44	0.439	69	5.475	146
Pr-19	21.4	8.28	0.91	583	241	0.455	292	78	24	0.236	38	2.944	79
Pr-20	21.4	8.29	0.76	485	201	0.379	243	65	20	0.196	31	2.449	65
Mean	**22.0**	**8.05**	**1.50**	**962**	**387**	**0.752**	**481**	**128.3**	**38.96**	**0.390**	**63.77**	**4.859**	**130**

Table 5.7: Physicochemical and microbiological characteristics of drinking water in *Patran* block of district Patiala

Table 5.8: Physicochemical and microbiological characteristics of drinking water in *Samana* block of district Patiala													
Sample	T °C	pH	EC dS/m	TDS mg/L	TA mg/L	TSS mg/L	TH mg/L	Ca^{2+} mg/L	Mg^{2+} mg/L	F^- mg/L	Cl^- mg/L	NO_3^- mg/L	Coliforms MPN/100ml
Sm-01	21.0	7.69	0.88	563	216	0.440	282	75	23	0.228	39	1.415	38
Sm-02	20.8	7.78	0.98	628	244	0.491	314	84	25	0.254	43	1.578	42
Sm-03	21.3	7.59	0.91	583	221	0.455	292	78	24	0.236	41	1.465	39
Sm-04	20.6	7.87	1.14	729	287	0.570	365	97	30	0.295	49	1.832	49
Sm-05	20.5	7.89	1.60	1026	405	0.802	513	137	42	0.416	69	2.578	69
Sm-06	20.8	7.78	1.47	938	365	0.733	469	125	38	0.380	64	2.357	63
Sm-07	21.3	7.58	0.93	593	225	0.463	297	79	24	0.240	42	1.490	40
Sm-08	21.3	7.58	0.83	529	200	0.413	265	71	21	0.214	37	1.329	35
Sm-09	20.9	7.75	0.68	435	169	0.340	218	58	18	0.176	30	1.093	29
Sm-10	20.6	7.87	0.71	456	179	0.356	228	61	18	0.185	31	1.146	31
Sm-11	21.6	7.48	1.15	739	276	0.577	370	99	30	0.299	53	1.857	50
Sm-12	20.6	7.84	1.53	982	385	0.767	491	131	40	0.398	67	2.467	66
Sm-13	21.0	7.69	1.71	1093	420	0.854	547	146	44	0.443	76	2.746	73
Sm-14	21.3	7.59	1.31	839	318	0.655	420	112	34	0.340	59	2.108	56
Sm-15	22.4	7.21	1.76	1128	407	0.881	564	150	46	0.457	83	2.834	76
Sm-16	21.8	7.43	1.97	1263	469	0.987	632	168	51	0.512	91	3.173	85
Sm-17	21.5	7.54	1.85	1184	446	0.925	592	158	48	0.480	84	2.975	79
Sm-18	21.8	7.42	1.31	839	311	0.655	420	112	34	0.340	60	2.108	56
Sm-19	22.4	7.23	1.30	829	300	0.648	415	111	34	0.336	61	2.083	56
Sm-20	22.7	7.14	1.14	729	260	0.570	365	97	30	0.295	54	1.832	49
Mean	**21.3**	**7.60**	**1.26**	**805**	**305**	**0.629**	**403**	**107.4**	**32.61**	**0.326**	**56.70**	**2.023**	**54**

Table 5.9: Physicochemical and microbiological characteristics of drinking water in *Sanour* block of district Patiala

Sample	T °C	pH	EC dS/m	TDS mg/L	TA mg/L	TSS mg/L	TH mg/L	Ca^{2+} mg/L	Mg^{2+} mg/L	F^- mg/L	Cl^- mg/L	NO_3^- mg/L	Coliforms MPN/100ml
Sr-01	22.0	8.03	1.78	1139	457	0.890	456	81	61	0.306	101	3.822	118
Sr-02	22.0	8.04	1.44	924	371	0.722	370	66	50	0.249	82	3.101	95
Sr-03	21.7	8.15	1.36	869	354	0.679	348	62	47	0.234	76	2.916	90
Sr-04	21.6	8.21	1.46	937	385	0.732	375	67	50	0.252	82	3.144	97
Sr-05	21.8	8.12	2.08	1329	540	1.038	532	95	72	0.358	117	4.460	137
Sr-06	21.9	8.09	1.30	829	335	0.648	332	59	45	0.223	73	2.782	86
Sr-07	21.6	8.19	2.07	1326	543	1.036	530	95	71	0.357	116	4.450	137
Sr-08	21.7	8.16	0.75	483	197	0.377	193	35	26	0.130	42	1.621	50
Sr-09	21.6	8.21	1.93	1238	508	0.967	495	88	67	0.333	108	4.154	128
Sr-10	21.9	8.08	1.45	927	375	0.724	371	66	50	0.249	82	3.111	96
Sr-11	22.1	8.01	1.99	1273	510	0.995	509	91	68	0.342	114	4.272	131
Sr-12	22.2	7.99	1.83	1173	469	0.916	469	84	63	0.316	105	3.936	121
Sr-13	21.9	8.08	1.15	739	299	0.577	296	53	40	0.199	65	2.480	76
Sr-14	22.0	8.04	1.52	973	391	0.760	389	70	52	0.262	86	3.265	100
Sr-15	21.7	8.17	1.47	938	383	0.733	375	67	50	0.252	82	3.148	97
Sr-16	21.6	8.21	0.98	629	258	0.491	252	45	34	0.169	55	2.111	65
Sr-17	21.4	8.26	1.76	1127	465	0.880	451	81	61	0.303	97	3.782	116
Sr-18	21.4	8.29	1.14	728	302	0.569	291	52	39	0.196	63	2.443	75
Sr-19	21.2	8.34	1.62	1038	433	0.811	415	74	56	0.279	89	3.483	107
Sr-20	21.4	8.26	1.30	829	342	0.648	332	59	45	0.223	72	2.782	86
Mean	**21.7**	**8.15**	**1.52**	**972**	**396**	**0.760**	**389**	**69.5**	**52.32**	**0.262**	**85.32**	**3.263**	**100**

Table 5.10: Comparison of Drinking Water Samples of *Bhunerheri* Block of district Patiala with BIS and WHO

Parameters	Range of Samples					BIS Standards IS:10500:1991		WHO Limit
	Min.	Max.	Mean	Standard Deviation	Variance	Desirable Limit	Permissible Limit	
Temperature, Odour, Taste and Appearance: Unobjectionable and Agreeable								
pH	7.72	8.26	8.00	0.144	0.021	6.5 – 8.5	6.5 – 8.5	7.0 – 8.5
EC	1.25	2.44	1.50	0.341	0.116	-	-	0.30
TDS	801	1563	987	218.005	47526.257	500	2000	500
TA	200	391	247	54.501	2970.391	200	600	100
TSS	0.559	0.589	0.576	0.008	5.859	-	-	-
TH	207	336	259	44.783	2005.566	300	600	100
Ca^{2+}	30	65	46.2	7.415	54.989	75	200	100
Mg^{2+}	22	63	34.7	12.740	162.295	30	75	50
F^-	0.199	1.030	0.500	0.240	0.058	1.00	1.50	1.00
Cl^-	86	117	99.7	6.794	46.152	250	1000	200
NO_3^-	2.013	3.927	2.500	0.548	0.300	45	100	45
Coliforms	81	157	99	21.910	480.051	Nil	-	Nil

Table 5.11: Comparison of Drinking Water Samples of *Ghanaur* Block of district Patiala with BIS and WHO

Parameters	Range of Samples					BIS Standards IS:10500:1991		WHO Limit
	Min.	Max.	Mean	Standard Deviation	Variance	Desirable Limit	Permissible Limit	
Temperature, Odour, Taste and Appearance: Unobjectionable and Agreeable								
pH	7.69	8.54	8.10	0.200	0.040	6.5 – 8.5	6.5 – 8.5	7.0 – 8.5
EC	0.73	3.70	1.30	0.631	0.398	-	-	0.30
TDS	467	2367	853	403.879	163118.041	500	2000	500
TA	263	496	319	70.708	4999.684	200	600	100
TSS	0.317	1.608	0.580	0.274	0.075	-	-	-
TH	159	504	278	101.669	10336.533	300	600	100
Ca^{2+}	16	74	28.8	12.918	166.873	75	200	100
Mg^{2+}	29	85	50	17.362	301.436	30	75	50
F^-	0.212	0.563	0.300	0.097	0.006	1.00	1.50	1.00
Cl^-	39	175	69.4	30.208	912.549	250	1000	200
NO_3^-	1.173	5.947	2.100	1.015	1.023	45	100	45
Coliforms	47	238	86	40.590	1647.615	Nil	-	Nil

Table 5.12: Comparison of Drinking Water Samples of *Nabha* Block of district Patiala with BIS and WHO

Parameters	Range of Samples					BIS Standards IS:10500:1991		WHO Limit
	Min.	Max.	Mean	Standard Deviation	Variance	Desirable Limit	Permissible Limit	
Temperature, Odour, Taste and Appearance: Unobjectionable and Agreeable								
pH	7.75	8.54	8.09	0.185	0.034	6.5 – 8.5	6.5 – 8.5	7.0 – 8.5
EC	0.69	2.24	1.12	0.470	0.220	-	-	0.30
TDS	439	1431	717	300.523	90314.006	500	2000	500
TA	218	492	293	80.258	6441.305	200	600	100
TSS	0.343	1.118	0.560	0.235	0.055	-	-	-
TH	220	516	339	112.876	12741.033	300	600	100
Ca^{2+}	22	72	35.9	15.026	225.785	75	200	100
Mg^{2+}	40	89	60.6	18.898	357.116	30	75	50
F^-	0.293	0.655	0.413	0.118	0.014	1.00	1.50	1.00
Cl^-	53	176	86.9	36.116	1304.364	250	1000	200
NO_3^-	1.103	3.595	1.802	0.755	0.570	45	100	45
Coliforms	22	72	36	15.102	228.060	Nil	-	Nil

Table 5.13: Comparison of Drinking Water Samples of *Rajpura* Block of district Patiala with BIS and WHO

Parameters	Range of Samples					BIS Standards IS:10500:1991		WHO Limit
	Min.	Max.	Mean	Standard Deviation	Variance	Desirable Limit	Permissible Limit	
Temperature, Odour, Taste and Appearance: Unobjectionable and Agreeable								
pH	6.98	8.17	7.64	0.421	0.178	6.5 – 8.5	6.5 – 8.5	7.0 – 8.5
EC	0.59	2.31	1.14	0.459	0.211	-	-	0.30
TDS	376	1476	726	293.834	86378.612	500	2000	500
TA	139	579	277	113.685	12924.314	200	600	100
TSS	0.294	1.153	0.567	0.230	0.053	-	-	-
TH	188	738	363	146.917	21524.653	300	600	100
Ca^{2+}	50	197	96.8	39.178	1534.909	75	200	100
Mg^{2+}	15	60	29.4	11.900	141.617	30	75	50
F^-	0.076	0.299	0.147	0.060	0.004	1.00	1.50	1.00
Cl^-	56	232	110.9	45.474	2067.890	250	1000	200
NO_3^-	3.837	15.061	7.407	2.999	8.990	45	100	45
Coliforms	102	402	198	79.955	6392.789	Nil	-	Nil

Table 5.14: Comparison of Drinking Water Samples of *Patiala* Block of district Patiala with BIS and WHO

Parameters	Range of Samples					BIS Standards IS:10500:1991		WHO Limit
	Min.	Max.	Mean	Standard Deviation	Variance	Desirable Limit	Permissible Limit	
Temperature, Odour, Taste and Appearance: Unobjectionable and Agreeable								
pH	6.95	8.42	7.80	0.434	0.188	6.5 – 8.5	6.5 – 8.5	7.0 – 8.5
EC	0.74	2.57	1.42	0.538	0.290	-	-	0.30
TDS	473	1643	911	344.366	118587.629	500	2000	500
TA	195	571	349	114.671	13149.486	200	600	100
TSS	0.370	1.284	0.712	0.269	0.072	-	-	-
TH	237	822	456	172.183	29646.907	300	600	100
Ca^{2+}	63	219	121.5	45.915	2108.224	75	200	100
Mg^{2+}	19	67	36.90	13.947	194.513	30	75	50
F^-	0.192	0.665	0.369	0.140	0.019	1.00	1.50	1.00
Cl^-	78	228	139.6	45.869	2103.918	250	1000	200
NO_3^-	1.587	5.513	3.057	1.156	1.335	45	100	45
Coliforms	42	147	82	30.816	949.608	Nil	-	Nil

Table 5.15: Comparison of Drinking Water Samples of *Patran* Block of district Patiala with BIS and WHO

Parameters	Range of Samples					BIS Standards IS:10500:1991		WHO Limit
	Min.	Max.	Mean	Standard Deviation	Variance	Desirable Limit	Permissible Limit	
Temperature, Odour, Taste and Appearance: Unobjectionable and Agreeable								
pH	7.86	8.36	8.05	0.161	0.026	6.5 – 8.5	6.5 – 8.5	7.0 – 8.5
EC	0.75	2.46	1.50	0.548	0.301	-	-	0.30
TDS	482	1572	962	350.925	123148.316	500	2000	500
TA	192	623	387	140.607	19770.411	200	600	100
TSS	0.377	1.228	0.752	0.274	0.075	-	-	-
TH	241	786	481	175.462	30787.079	300	600	100
Ca^{2+}	64	210	128.3	46.790	2189.303	75	200	100
Mg^{2+}	20	64	38.96	14.212	201.994	30	75	50
F^-	0.195	0.637	0.390	0.142	0.020	1.00	1.50	1.00
Cl^-	31	106	63.77	23.443	549.581	250	1000	200
NO_3^-	2.434	7.939	4.859	1.772	3.141	45	100	45
Coliforms	65	212	130	47.262	2233.755	Nil	-	Nil

Table 5.16: Comparison of Drinking Water Samples of *Samana* Block of district Patiala with BIS and WHO

Parameters	Range of Samples					BIS Standards IS:10500:1991		WHO Limit
	Min.	Max.	Mean	Standard Deviation	Variance	Desirable Limit	Permissible Limit	
Temperature, Odour, Taste and Appearance: Unobjectionable and Agreeable								
pH	7.14	7.89	7.60	0.226	0.051	6.5 – 8.5	6.5 – 8.5	7.0 – 8.5
EC	0.68	1.97	1.26	0.389	0.151	-	-	0.30
TDS	435	1263	805	248.821	61912.052	500	2000	500
TA	169	469	305	92.954	8640.483	200	600	100
TSS	0.340	0.987	0.629	0.194	0.038	-	-	-
TH	218	632	403	124.411	15448.023	300	600	100
Ca^{2+}	58	168	107.4	33.176	1100.660	75	200	100
Mg^{2+}	18	51	32.61	10.077	101.551	30	75	50
F^-	0.176	0.512	0.326	0.101	0.010	1.00	1.50	1.00
Cl^-	30	91	56.70	17.960	322.537	250	1000	200
NO_3^-	1.093	3.173	2.023	0.625	0.391	45	100	45
Coliforms	29	85	54	16.671	277.937	Nil	-	Nil

Table 5.17: Comparison of Drinking Water Samples of *Sanour* Block of district Patiala with BIS and WHO

Parameters	Range of Samples					BIS Standards IS:10500:1991		WHO Limit
	Min.	Max.	Mean	Standard Deviation	Variance	Desirable Limit	Permissible Limit	
Temperature, Odour, Taste and Appearance: Unobjectionable and Agreeable								
pH	7.99	8.34	8.15	0.100	0.010	6.5 – 8.5	6.5 – 8.5	7.0 – 8.5
EC	0.75	2.08	1.52	0.365	0.134	-	-	0.30
TDS	483	1329	972	233.914	54715.937	500	2000	500
TA	197	543	396	94.483	8927.038	200	600	100
TSS	0.377	1.038	0.760	0.183	0.033	-	-	-
TH	193	532	389	93.566	8754.550	300	600	100
Ca^{2+}	35	95	69.5	16.708	279.163	75	200	100
Mg^{2+}	26	72	52.32	12.586	158.414	30	75	50
F^-	0.130	0.358	0.262	0.063	0.004	1.00	1.50	1.00
Cl^-	42	117	85.32	20.743	430.302	250	1000	200
NO_3^-	1.621	4.460	3.263	0.785	0.616	45	100	45
Coliforms	50	137	100	24.152	583.330	Nil	-	Nil

Table 5.18: Classification of *Drinking Water* in Villages of different Blocks of district Patiala on the basis of Total Hardness
(Scale based on Matalas and Reiher, 1967 & Sawyer and McCarthy, 1967)

TH Mg/L	Description	Number of Samples Block-wise							
		Bhunerheri	Ghanaur	Nabha	Patran	Patiala	Rajpura	Samana	Sanour
< 75	Soft	0	0	0	0	0	0	0	0
75-150	Moderately Hard	0	0	0	0	0	0	0	0
151-300	Hard	24	20	15	05	05	13	06	03
> 300	Very Hard	06	10	15	15	15	17	14	17
	TOTAL	30	30	30	20	20	30	20	20

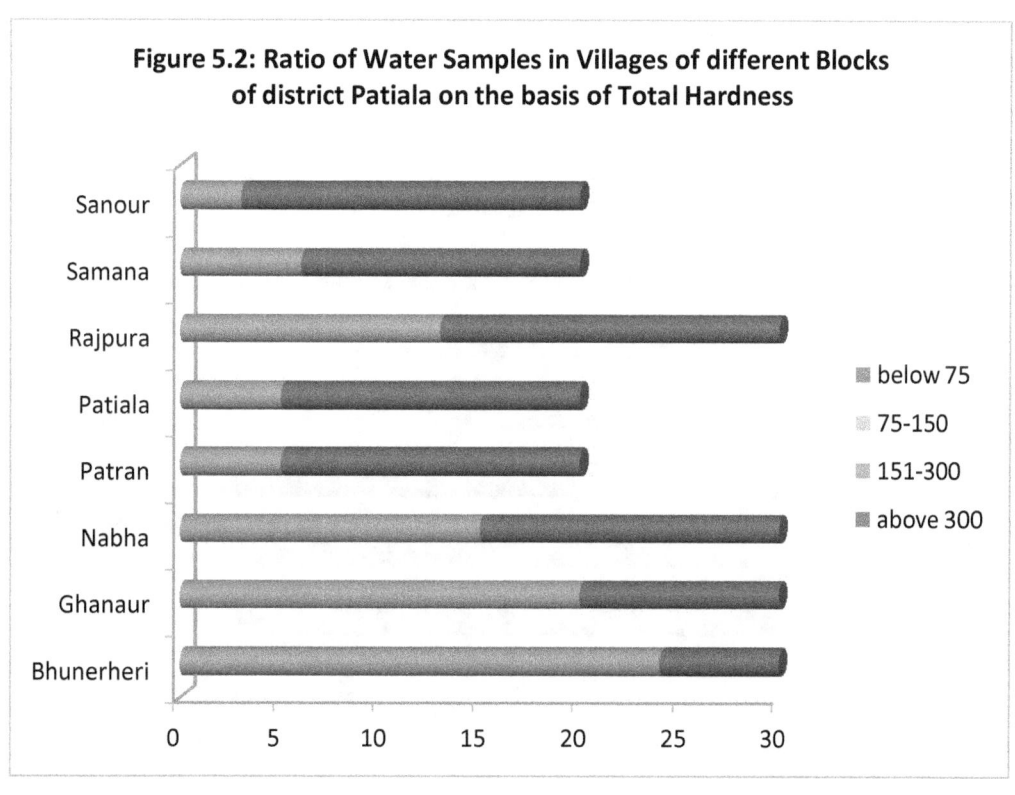

Figure 5.2: Ratio of Water Samples in Villages of different Blocks of district Patiala on the basis of Total Hardness

TDS Mg/L	Description	Number of Samples Block-wise							
		Bhunerheri	Ghanaur	Nabha	Patran	Patiala	Rajpura	Samana	Sanour
< 300	Good	0	0	0	0	0	0	0	0
300-500	Fair	0	05	13	03	04	10	02	01
500-900	Average	18	16	09	04	06	12	11	06
900-1200	Poor	06	05	04	08	06	05	06	09
1200-2000	Very Poor	06	03	04	05	04	03	01	04
>2000	Unacceptable	0	01	0	0	0	0	0	0
	TOTAL	30	30	30	20	20	30	20	20

Table 5.19: Classification of *Drinking Water* in Villages of different Blocks of district Patiala on the basis of Total Dissolved Solids (Scale based on Matalas and Reiher, 1967)

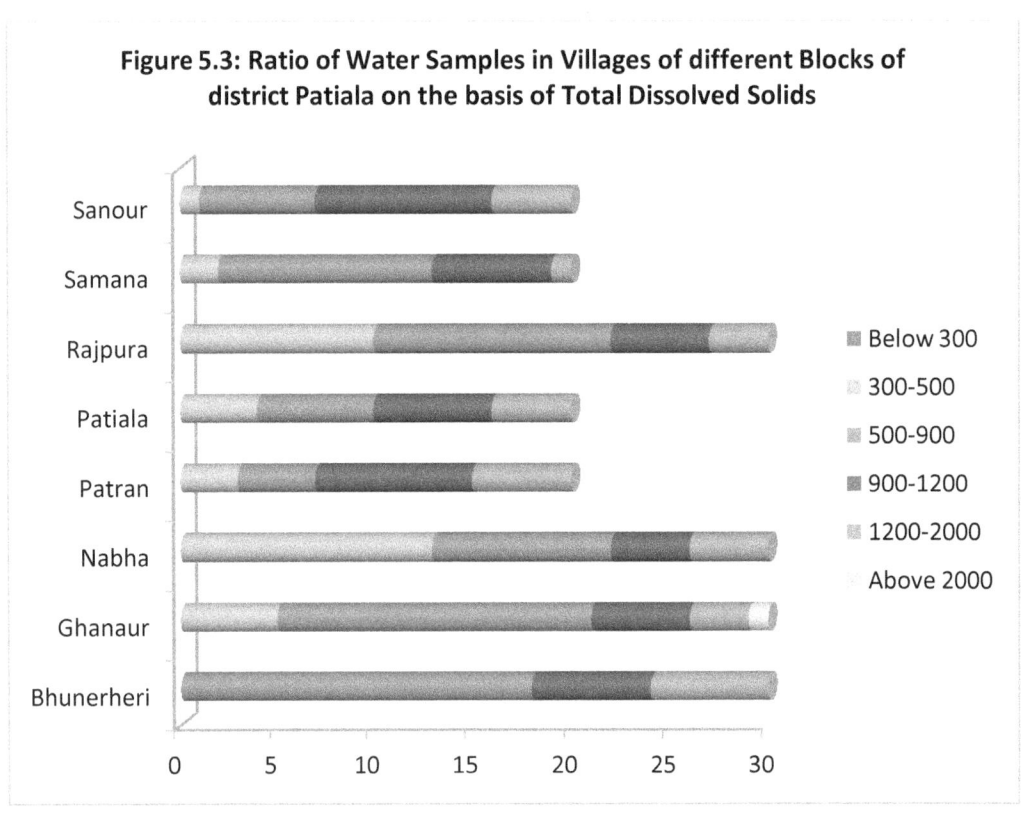

Figure 5.3: Ratio of Water Samples in Villages of different Blocks of district Patiala on the basis of Total Dissolved Solids

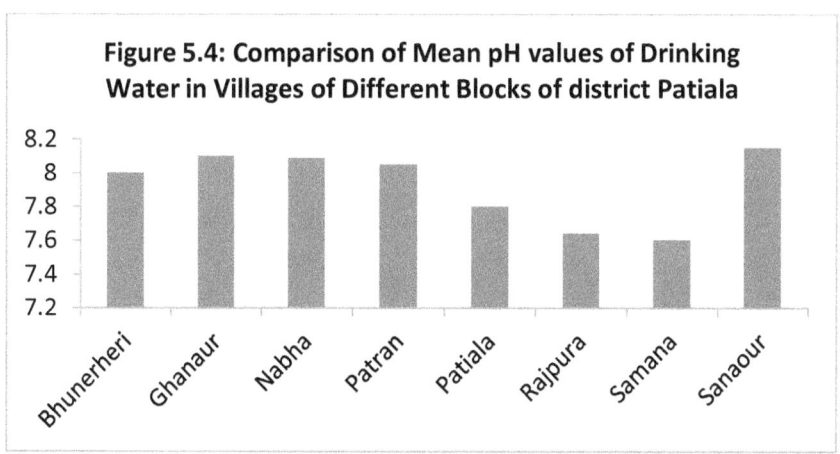

Figure 5.4: Comparison of Mean pH values of Drinking Water in Villages of Different Blocks of district Patiala

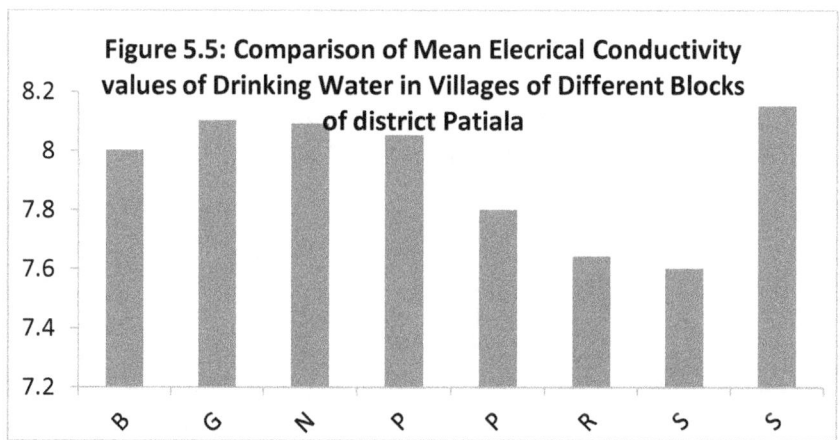

Figure 5.5: Comparison of Mean Elecrical Conductivity values of Drinking Water in Villages of Different Blocks of district Patiala

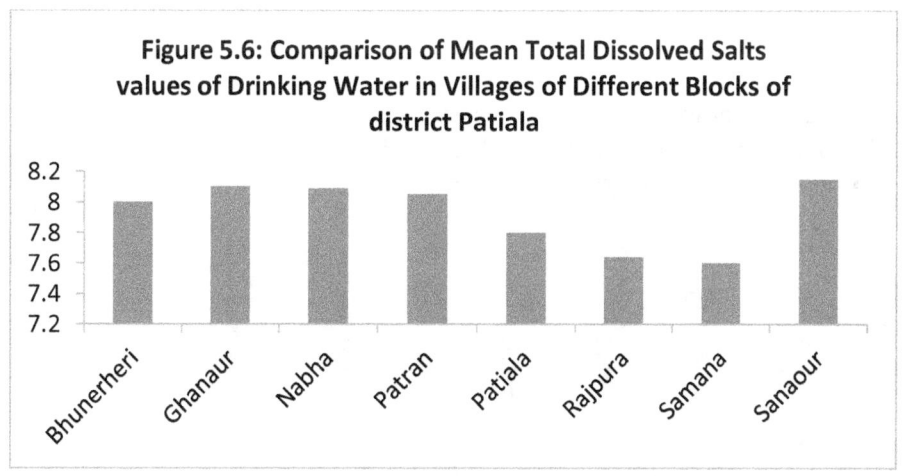

Figure 5.6: Comparison of Mean Total Dissolved Salts values of Drinking Water in Villages of Different Blocks of district Patiala

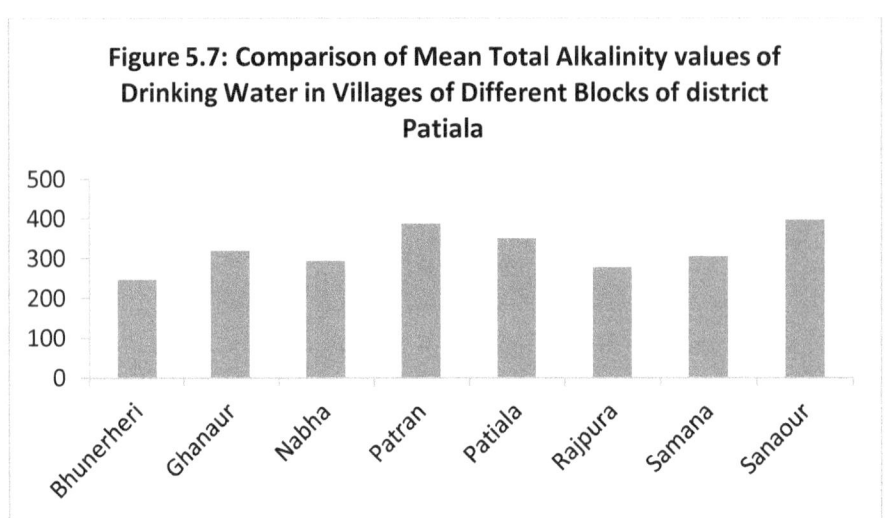

Figure 5.7: Comparison of Mean Total Alkalinity values of Drinking Water in Villages of Different Blocks of district Patiala

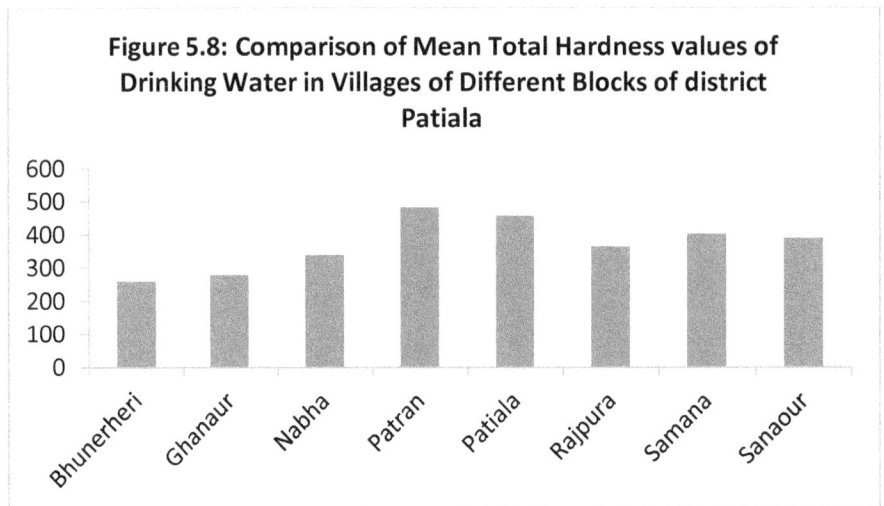

Figure 5.8: Comparison of Mean Total Hardness values of Drinking Water in Villages of Different Blocks of district Patiala

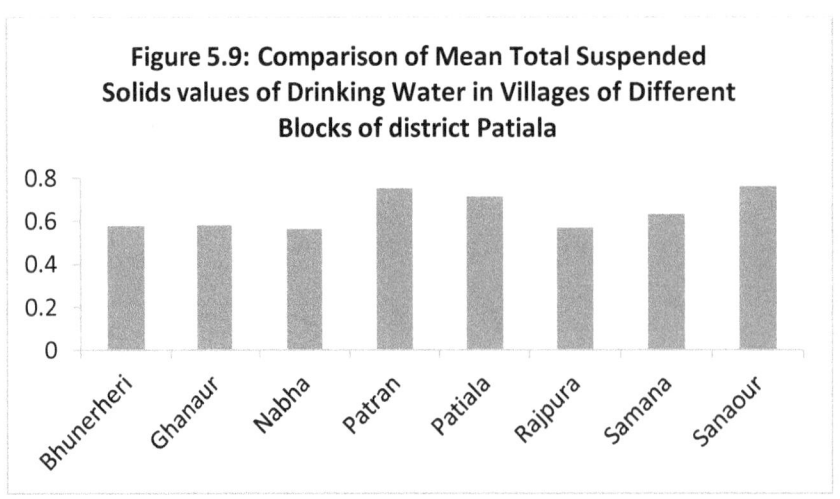

Figure 5.9: Comparison of Mean Total Suspended Solids values of Drinking Water in Villages of Different Blocks of district Patiala

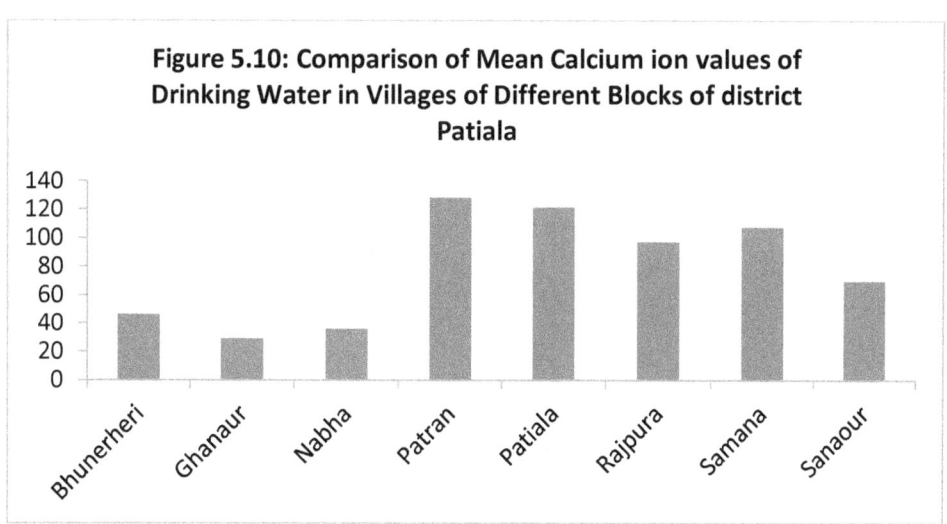

Figure 5.10: Comparison of Mean Calcium ion values of Drinking Water in Villages of Different Blocks of district Patiala

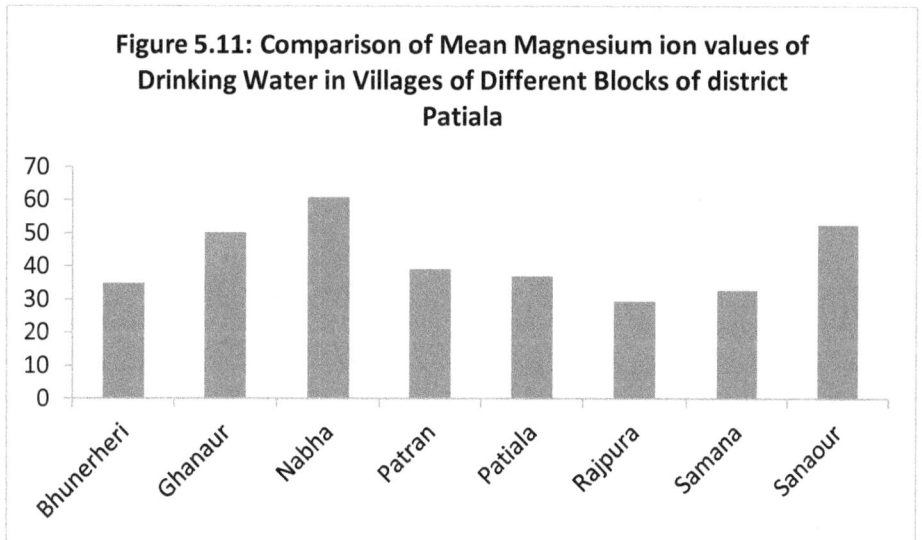

Figure 5.11: Comparison of Mean Magnesium ion values of Drinking Water in Villages of Different Blocks of district Patiala

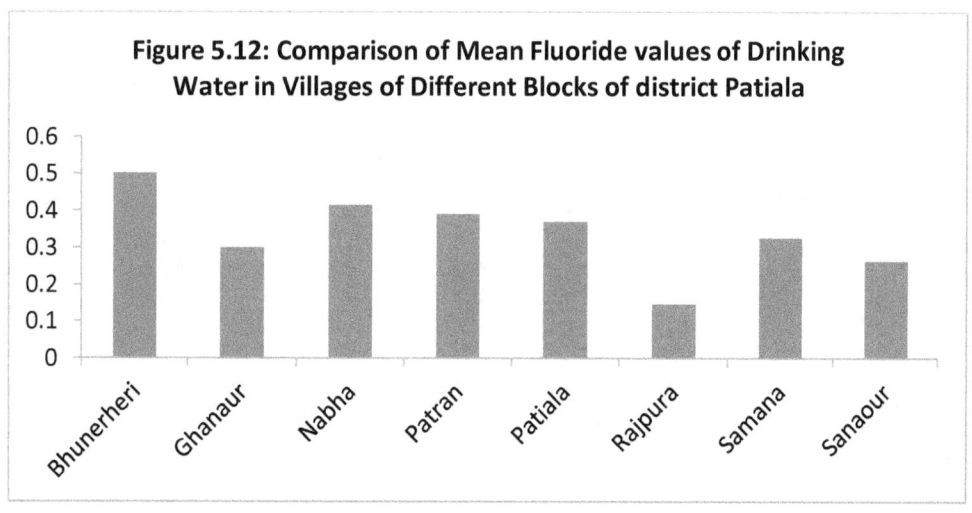

Figure 5.12: Comparison of Mean Fluoride values of Drinking Water in Villages of Different Blocks of district Patiala

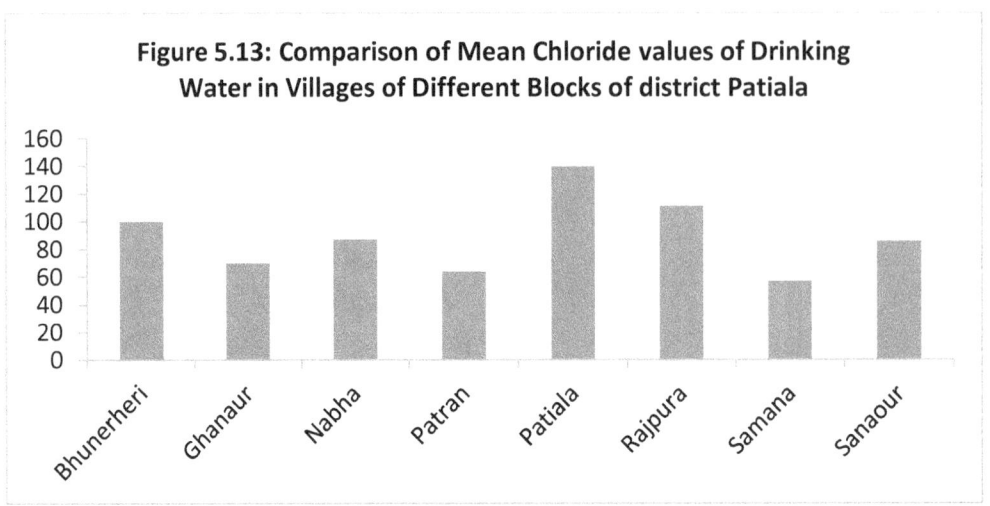

Figure 5.13: Comparison of Mean Chloride values of Drinking Water in Villages of Different Blocks of district Patiala

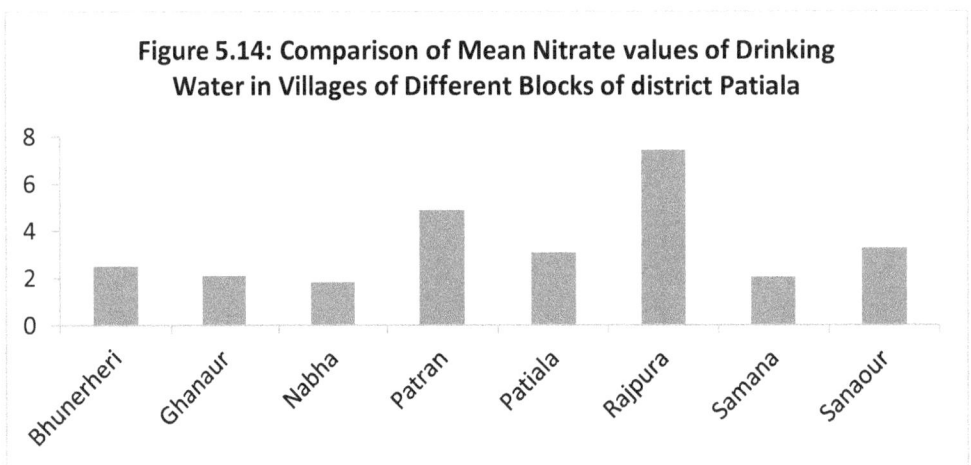

Figure 5.14: Comparison of Mean Nitrate values of Drinking Water in Villages of Different Blocks of district Patiala

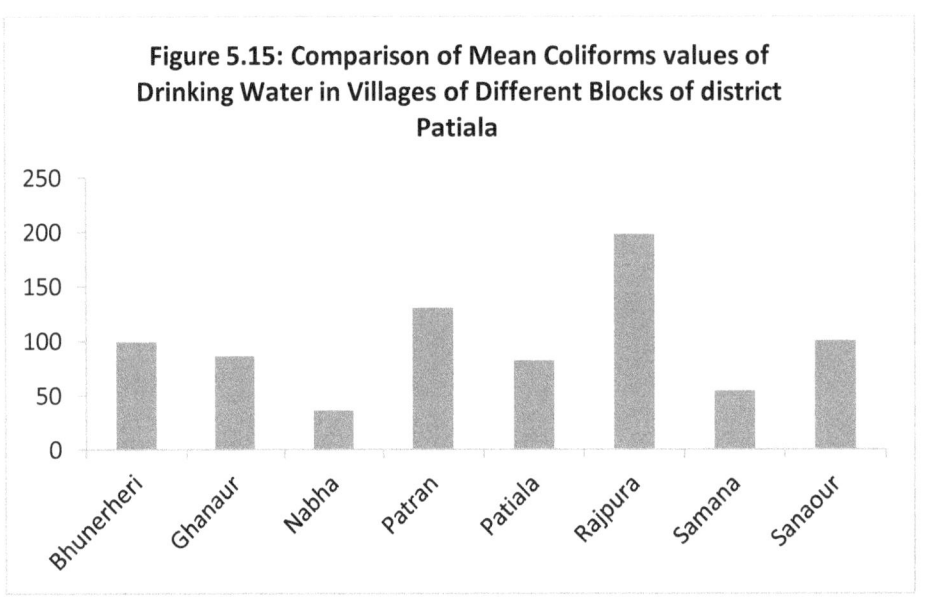

Figure 5.15: Comparison of Mean Coliforms values of Drinking Water in Villages of Different Blocks of district Patiala

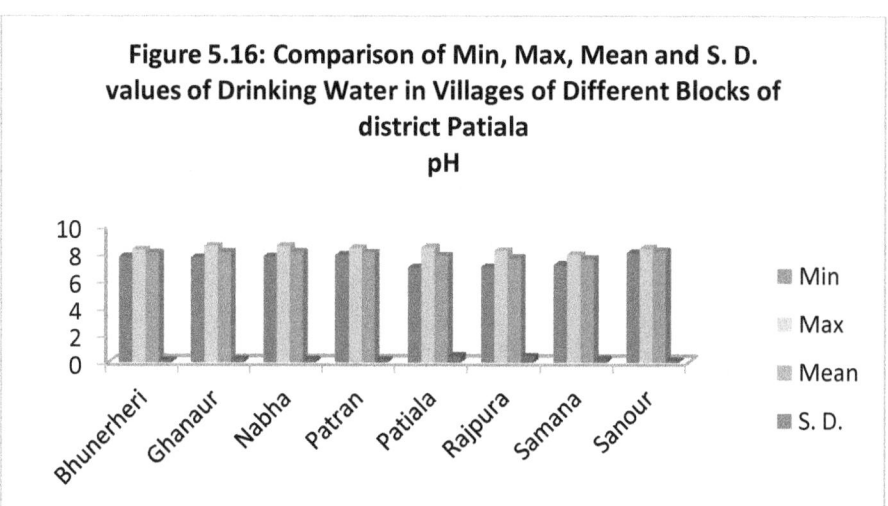

Figure 5.16: Comparison of Min, Max, Mean and S. D. values of Drinking Water in Villages of Different Blocks of district Patiala
pH

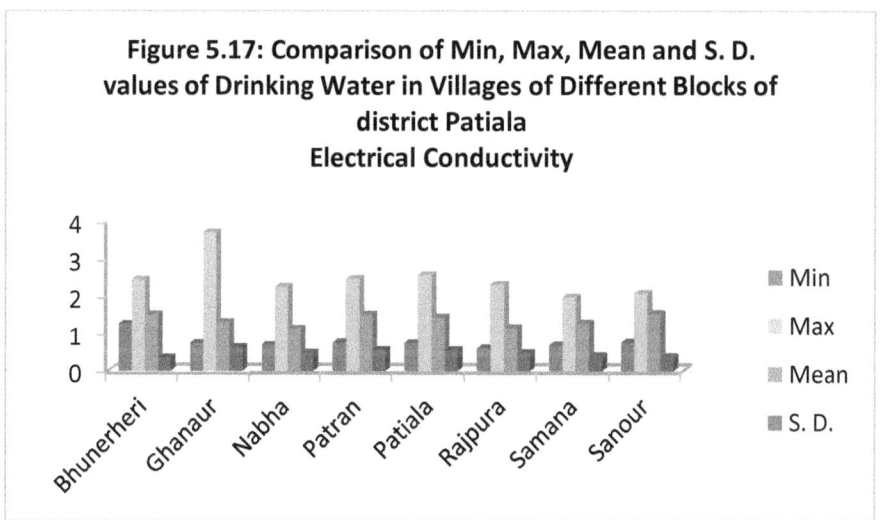

Figure 5.17: Comparison of Min, Max, Mean and S. D. values of Drinking Water in Villages of Different Blocks of district Patiala
Electrical Conductivity

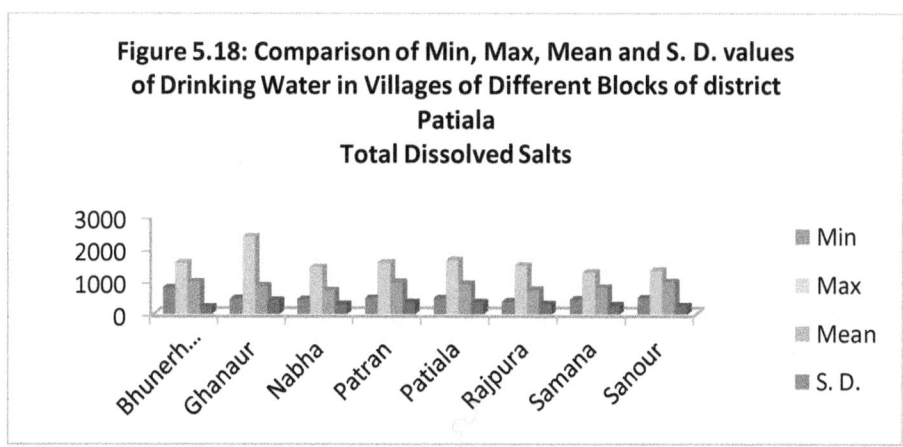

Figure 5.18: Comparison of Min, Max, Mean and S. D. values of Drinking Water in Villages of Different Blocks of district Patiala
Total Dissolved Salts

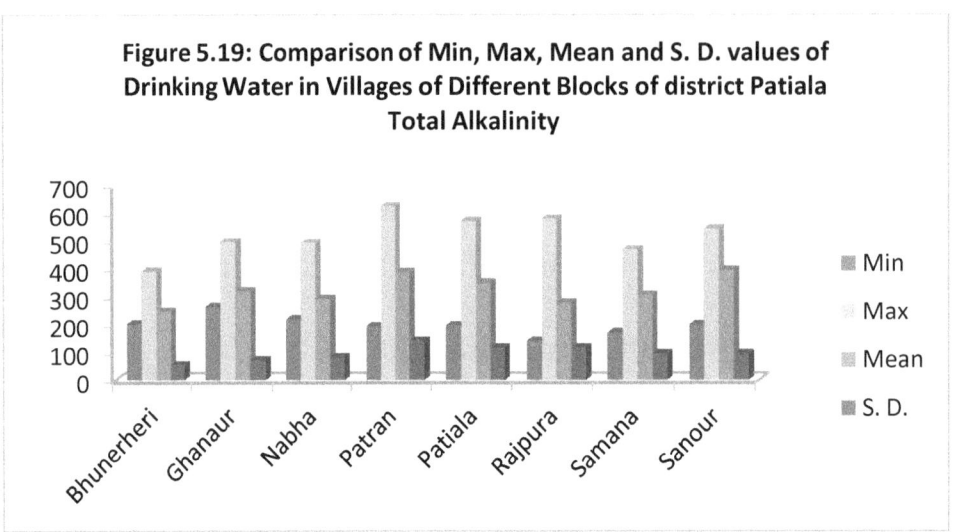

Figure 5.19: Comparison of Min, Max, Mean and S. D. values of Drinking Water in Villages of Different Blocks of district Patiala Total Alkalinity

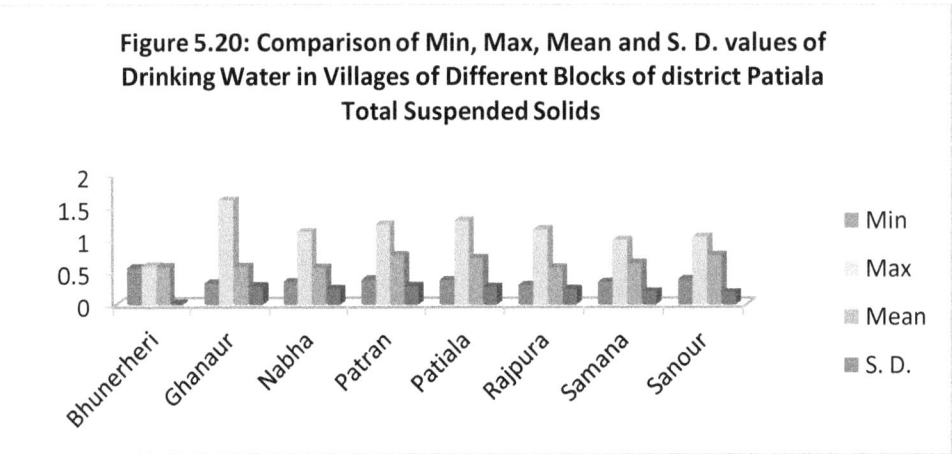

Figure 5.20: Comparison of Min, Max, Mean and S. D. values of Drinking Water in Villages of Different Blocks of district Patiala Total Suspended Solids

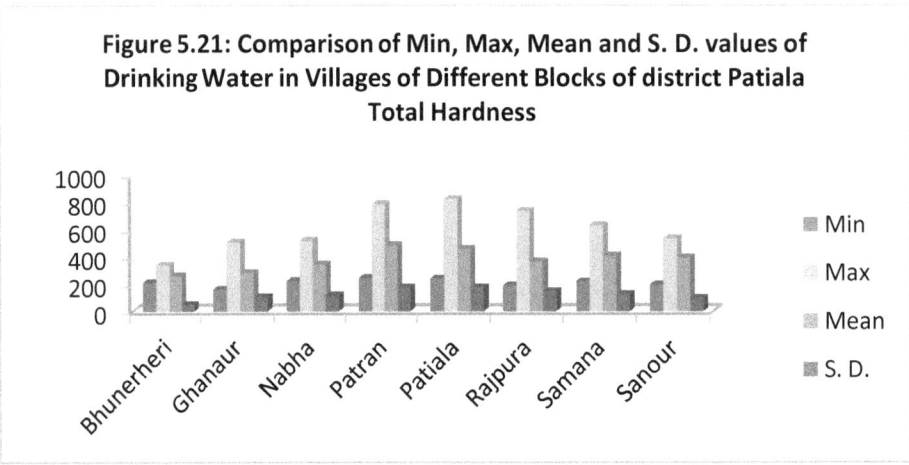

Figure 5.21: Comparison of Min, Max, Mean and S. D. values of Drinking Water in Villages of Different Blocks of district Patiala Total Hardness

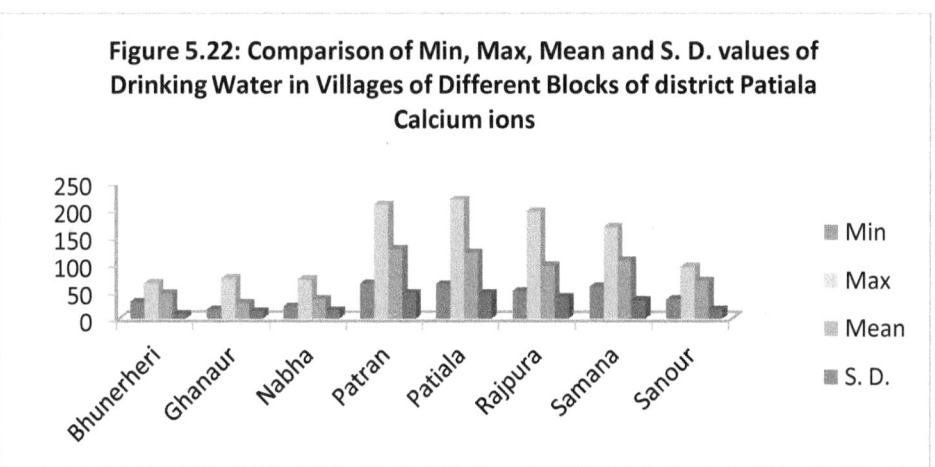

Figure 5.22: Comparison of Min, Max, Mean and S. D. values of Drinking Water in Villages of Different Blocks of district Patiala Calcium ions

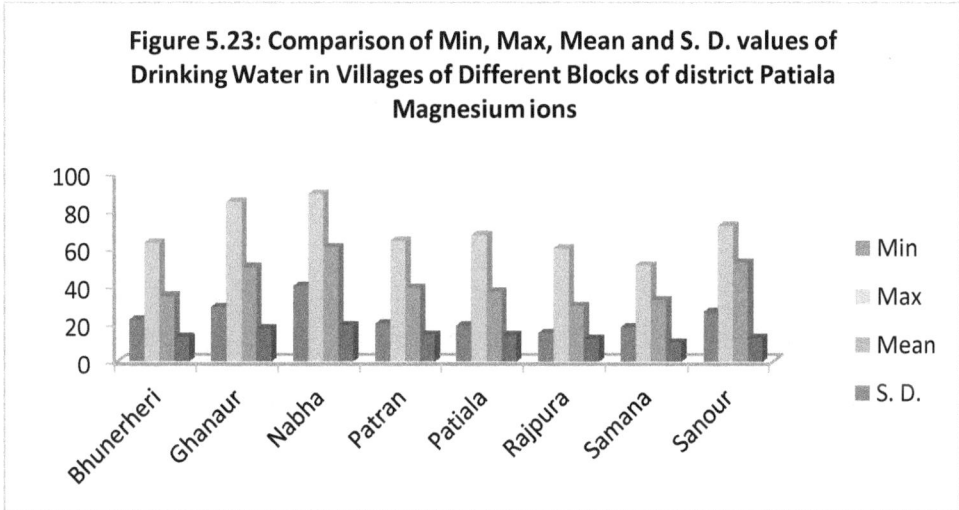

Figure 5.23: Comparison of Min, Max, Mean and S. D. values of Drinking Water in Villages of Different Blocks of district Patiala Magnesium ions

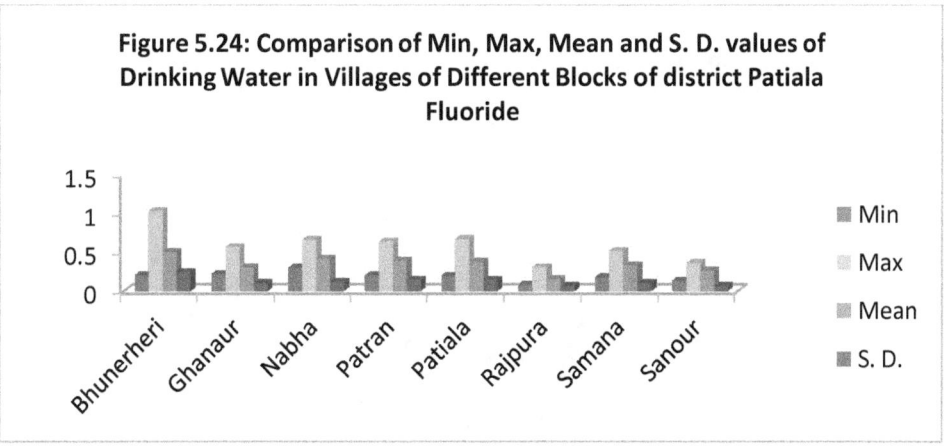

Figure 5.24: Comparison of Min, Max, Mean and S. D. values of Drinking Water in Villages of Different Blocks of district Patiala Fluoride

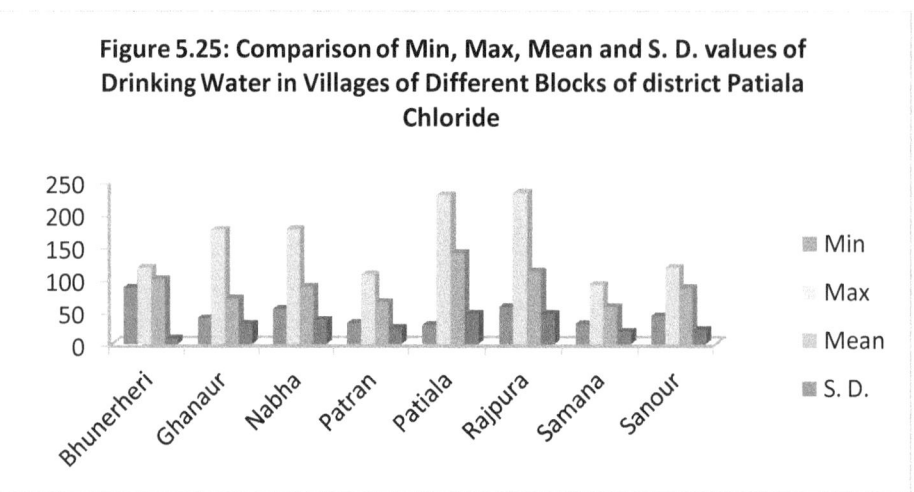

Figure 5.25: Comparison of Min, Max, Mean and S. D. values of Drinking Water in Villages of Different Blocks of district Patiala Chloride

Figure 5.26: Comparison of Min, Max, Mean and S. D. values of Drinking Water in Villages of Different Blocks of district Patiala Nitrate

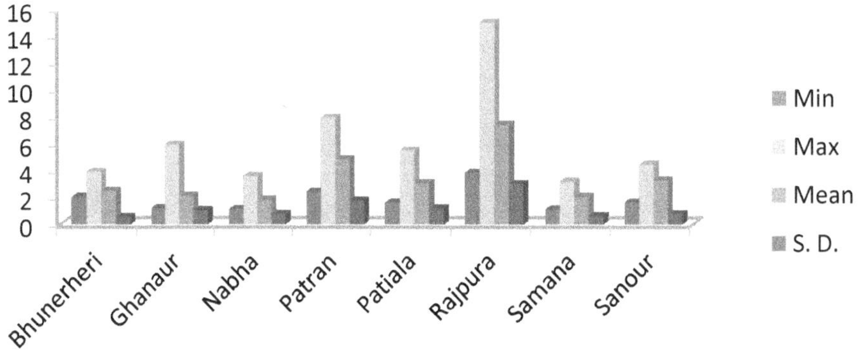

Figure 5.27: Comparison of Min, Max, Mean and S. D. values of Drinking Water in Villages of Different Blocks of district Patiala Coliforms

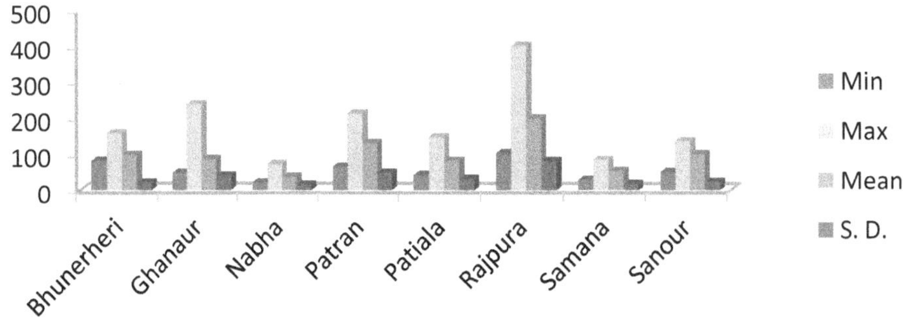

Table 5.20: Mean Values of different parameters of drinking water of the blocks where Ghaggar River passes				
Parameter	Mean values of the villages fall under blocks		Mean values of Group - A	Standard Deviation
	Bhunerheri	Ghanaur		
pH	8.00	8.10	8.05	0.0707
EC	1.50	1.30	1.40	0.1414
TDS	987	853	920	94.7523
TA	247	319	283	50.9117
TSS	0.576	0.580	0.578	0.0028
TH	259	278	269	13.4350
Ca^{2+}	46.2	28.8	37.5	12.3037
Mg^{2+}	34.7	50.0	42.4	10.8187
F^-	0.500	0.300	0.400	0.1414
Cl^-	99.70	69.40	84.55	21.4253
NO_3^-	2.500	2.100	2.3	0.2828
Coliforms	99	86	93	9.1924

Table 5.21: Mean Values of different parameters of drinking water of the blocks where Patiale-wali Nadi passes				
Parameter	Mean values of the villages fall under blocks		Mean values of Group - B	Standard Deviation
	Patiala	Sanour		
pH	7.8	8.15	8.0	0.2475
EC	1.42	1.52	1.47	0.0707
TDS	912	972	942	42.4264
TA	349	396	373	33.2340
TSS	0.712	0.760	0.736	0.0339
TH	456	389	423	47.3761
Ca^{2+}	121.5	69.5	95.5	36.7695
Mg^{2+}	36.9	52.32	44.61	10.9036
F^-	0.369	0.262	0.316	0.0757
Cl^-	139.60	85.32	112.46	38.3818
NO_3^-	3.057	3.263	3.160	0.1457
Coliforms	82	100	91	12.7279

Table 5.22: Mean Values of different parameters of drinking water of the blocks where Ghaggar River and Patiale-wali Nadi passes (Cluster – 1)

Parameter	Mean values of the villages fall under blocks					Mean values of Cluster - 1	Standard Deviation
	Bhunerheri	Ghanaur	Patiala	Patran	Sanour		
pH	8.00	8.10	7.8	8.05	8.15	**8.02**	0.1351
EC	1.50	1.30	1.42	1.50	1.52	**1.49**	0.0912
TDS	987	853	912	962	972	**937**	54.8425
TA	247	319	349	387	396	**340**	60.2395
TSS	0.576	0.580	0.712	0.752	0.760	**0.676**	0.0913
TH	259	278	456	481	389	**373**	101.0312
Ca^{2+}	46.2	28.8	121.5	128.3	69.5	**78.9**	44.5049
Mg^{2+}	34.7	50.0	36.9	38.96	52.32	**42.58**	8.0216
F^-	0.500	0.300	0.369	0.390	0.262	**0.364**	0.0918
Cl^-	99.70	69.40	139.6	63.77	85.32	**91.56**	30.3166
NO_3^-	2.500	2.100	3.057	4.859	3.263	**3.156**	1.0568
Coliforms	99	86	82	130	100	**100**	18.8361

	pH	EC	TDS	TA	TSS	TH	Ca^{2+}	Mg^{2+}	F$^-$	Cl$^-$	NO$_3^-$	Coliforms
Table 5.23: Correlation Coefficients among different drinking Water Quality Parameters of the villages studied of the district Patiala (CLUSTER-1)												
pH	1.0000											
EC	0.0852	1.0000										
TDS	0.1039	0.9771	1.0000									
TA	0.2117	0.2046	0.0018	1.0000								
TSS	-0.0081	0.5403	0.3497	0.9053	1.0000							
TH	-0.3345	0.3412	0.1564	0.7934	0.8960	1.0000						
Ca^{2+}	-0.5184	0.3950	0.2473	0.5849	0.7739	0.9568	1.0000					
Mg^{2+}	0.7286	-0.2856	-0.3551	0.4662	0.1414	-0.1583	-0.4387	1.0000				
F$^-$	-0.4319	0.3109	0.4535	-0.7121	-0.4137	-0.2112	0.0744	-0.9041	1.0000			
Cl$^-$	-0.8637	0.0419	0.0232	-0.2074	0.0367	0.1684	0.2921	-0.4745	0.2407	1.0000		
NO$_3^-$	0.0270	0.5737	0.4516	0.6610	0.7821	0.8255	0.8147	-0.2131	0.0290	-0.2707	1.0000	
Coliforms	0.3694	0.5943	0.5631	0.3600	0.4472	0.4189	0.4322	-0.1694	0.2098	-0.6113	0.8580	1.0000

Parameter	Mean values of the villages fall under blocks			Mean values of Cluster - 2	Standard Deviation
	Nabha	Rajpura	Samana		
pH	8.09	7.64	7.6	**7.78**	0.2720
EC	1.12	1.14	1.26	**1.17**	0.0757
TDS	717	726	805	**749**	48.4183
TA	293	277	305	**292**	14.0475
TSS	0.56	0.567	0.629	**0.585**	0.0380
TH	339	363	403	**368**	32.3316
Ca^{2+}	35.9	96.8	107.4	**80.03**	38.5863
Mg^{2+}	60.6	29.4	32.61	**40.87**	17.1619
F^-	0.413	0.147	0.326	**0.295**	0.1356
Cl^-	86.9	110.9	56.7	**84.83**	27.1590
NO_3^-	1.802	7.407	2.023	**3.744**	3.1742
Coliforms	36	198	54	**96**	88.7919

Table 5.24: Mean Values of different parameters of drinking water of Other Blocks (Cluster – 2)

Table 5.25: Correlation Coefficients among different drinking Water Quality Parameters of the villages studied of the district Patiala (CLUSTER-2)

	pH	EC	TDS	TA	TSS	TH	Ca^{2+}	Mg^{2+}	F$^-$	Cl$^-$	NO$_3^-$	Coliforms
pH	1.0000											
EC	-0.6666	1.0000										
TDS	-0.6367	0.9992	1.0000									
TA	0.0087	0.7396	0.7655	1.0000								
TSS	-0.6361	0.9992	1.0000	0.7660	1.0000							
TH	-0.8291	0.9695	0.9590	0.5519	0.9588	1.0000						
Ca^{2+}	-0.9979	0.7131	0.6849	0.0555	0.6843	0.8632	1.0000					
Mg^{2+}	0.9860	-0.5332	-0.4995	0.1750	-0.4988	-0.7244	-0.9733	1.0000				
F$^-$	0.7008	0.0646	0.1038	0.7194	0.1046	-0.1821	-0.6536	0.8098	1.0000			
Cl$^-$	0.1391	-0.8309	-0.8521	-0.9890	-0.8526	-0.6690	-0.2023	-0.0277	-0.6090	1.0000		
NO$_3^-$	-0.4661	-0.3488	-0.3855	-0.8888	-0.3862	-0.1083	0.4083	-0.6068	-0.9578	0.8113	1.0000	
Coliforms	-0.5240	-0.2856	-0.3231	-0.8562	-0.3238	-0.0418	0.4683	-0.6585	-0.9748	0.7706	0.9978	1.0000

Figure 5.28: Comparison of different water parameters of Cluster-1 and Cluster-2

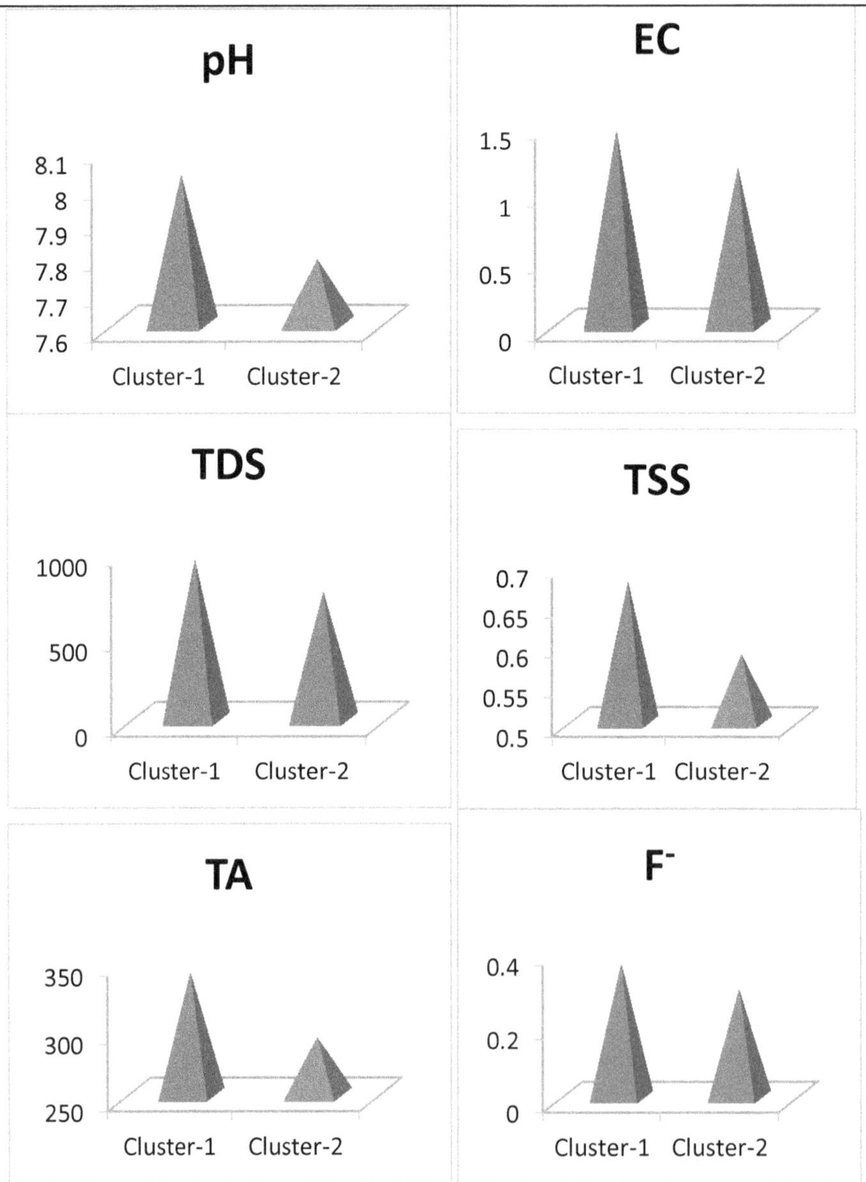

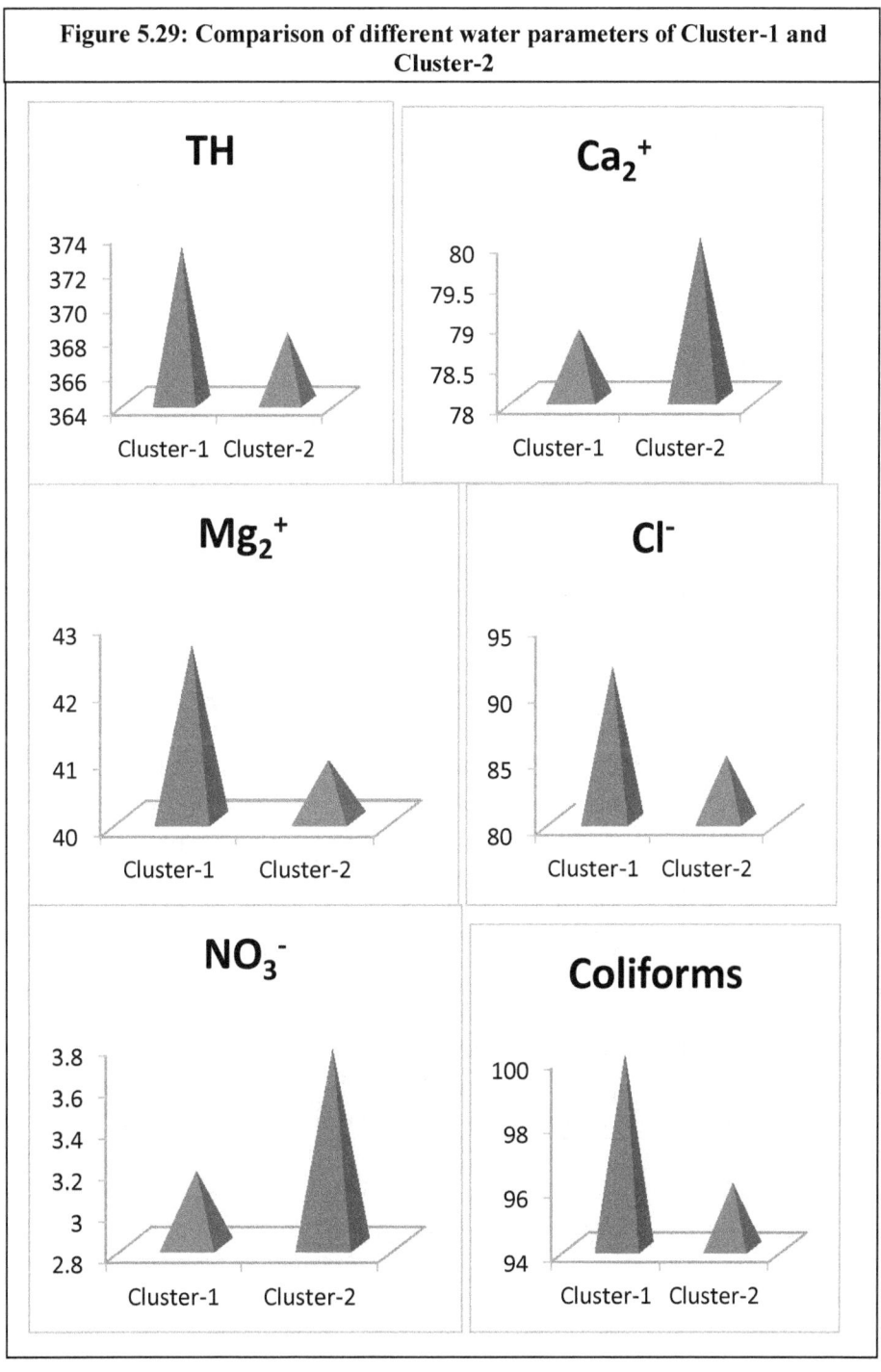

Figure 5.29: Comparison of different water parameters of Cluster-1 and Cluster-2

Table 5.26: Mean Values of different parameters of drinking water of villages studied of the district Patiala										
Parameter	Mean Values of different Parameters in Villages studied (Block-wise)								Mean Values of the district Patiala	Standard Deviation
	Bhunerheri	Ghanaur	Nabha	Patran	Patiala	Rajpura	Samana	Sanour		
pH	8.00	8.10	8.09	8.05	7.80	7.64	7.60	8.15	**7.93**	0.2178
EC	1.50	1.30	1.12	1.50	1.42	1.14	1.26	1.52	**1.345**	0.1631
TDS	987	853	717	962	912	726	805	972	**867**	108.8219
TA	247	319	293	387	349	277	305	396	**322**	52.3967
TSS	0.576	0.580	0.560	0.752	0.712	0.567	0.629	0.760	**0.642**	0.0859
TH	259	278	339	481	456	363	403	389	**371**	78.3345
Ca^{2+}	46.2	28.8	35.9	128.3	121.5	96.8	107.4	69.5	**79.3**	39.4664
Mg^{2+}	34.7	50.0	60.6	38.96	36.90	29.40	32.61	52.32	**41.94**	11.0318
F^-	0.500	0.300	0.413	0.390	0.369	0.147	0.326	0.262	**0.338**	0.1065
Cl^-	99.70	69.40	86.90	63.77	139.60	110.90	56.70	85.32	**89.04**	27.3507
NO_3^-	2.500	2.100	1.802	4.859	3.057	7.407	2.023	3.263	**3.376**	1.8999
Coliforms	99	86	36	130	82	198	54	100	**98**	49.5824
Units of all the parameters are mg/L except EC (dS/m) and Total Coliforms (MPN/100ml) and pH										

Table 5.27: Correlation Coefficients among different drinking Water Quality Parameters of all the villages studied of the district Patiala											
	pH	EC	TDS	TA	TSS	TH	Ca^{2+}	Mg^{2+}	F^-	Cl^-	NO_3^-	Coliforms
pH	1.0000											
EC	0.4104	1.0000										
TDS	0.4341	0.9951	1.0000									
TA	0.3608	0.5141	0.4497	1.0000								
TSS	0.2124	0.7183	0.6514	0.9168	1.0000							
TH	-0.2587	0.2183	0.1336	0.7031	0.7673	1.0000						
Ca^{2+}	-0.5643	0.2214	0.1517	0.4302	0.6062	0.8942	1.0000					
Mg^{2+}	0.7816	-0.1066	-0.1016	0.2815	0.0081	-0.2155	-0.6298	1.0000				
F^-	0.3802	0.3883	0.4284	-0.1746	-0.0169	-0.1521	-0.1964	0.1614	1.0000			
Cl^-	-0.2164	0.0163	0.0135	-0.1660	-0.0127	0.0628	0.1505	-0.2206	-0.0460	1.0000		
NO_3^-	-0.3653	-0.1150	-0.1527	0.0520	0.0952	0.3125	0.4850	-0.5127	-0.6280	0.2688	1.0000	
Coliforms	-0.2647	0.0352	0.0198	-0.0108	0.0493	0.1095	0.3395	-0.5480	-0.5841	0.2489	0.9509	1.0000

Chapter-6

RESULTS AND DISCUSSION

The quality of drinking water is a powerful environmental determinant of health. The contamination of natural water with domestic/ industrial/ agricultural/ human/ livestock waste and pasture runoff may result in an increased risk of disease transmission to humans (Geldreich, 1991). The major causes of water pollution are discharges of untreated domestic sewage and untreated industrial wastes in to the surface water bodies and on land, increasing use of fertilizer, pesticides and dumping of organic and inorganic wastes in to water. Diarrheal disease from contaminated water continues to be a serious problem in developing countries and also in developed countries (Grant, 1997). In quantitative terms, only one percent world water source is available for different human uses (AWWA 1999). Once viewed as an infinite and bountiful resource, water today defines human, social, and economic development. Present scenario indexing towards water crisis. 85% of rural India depends on groundwater and groundwater is depleting at an alarming rate. Overexploitation is not only emptying the aquifers but also contaminating the water. The shortage and contamination of water is slowly affecting the lives of people as well as the environment around them.

Water required for public use must be potable i.e., satisfactory for drinking purposes from the standpoint of its chemical, physical and biological characteristics. Drinking water should, preferably, be obtained from a source free from pollution. A daily per capita consumption of 2 liters is generally accepted value for a person weighing 60 Kg. This is the value used in estimating ingestion exposure to potentially hazardous chemicals in drinking water. The actual water intake, however, varies from individual to individual and according to climate, physical activity and culture. Water need increases sharply as ambient temperature exceeds 25°C, primarily to make up for moisture loss through perspiration. Infants and children consume more water per unit weight than adults.

From ancient time onwards history has spoken a lot on the role and importance of water in religious, cultural and socioeconomic traditions. Approach to surface water for everyone and anytime, is not feasible and easy because it is not uniformly and evenly distributed. So, the groundwater became an important source for existence and economic development. Moreover, groundwater is considered safe than surface water because it is away from direct contamination by environmental and humanized factors.

Due to less availability and non-acceptance of surface water, people of Punjab and also Patiala, have to depend on groundwater resources to a greater extent. Even, in the many areas, groundwater is the main and only source of water. In rural areas, wells, hand-pumps, submersible pumps and tube-wells serve the main and fresh water source of drinking water. Governmental water supply is available for limited times in urban areas, just three hours each time in the morning and evening and one hour in afternoon. In rural areas, where governmental water supplies are available, there is no fixed time or duration. Less availability of electricity, also affects the supply of water. Sometimes once a day, after two-three days, or may be once a week, or any other time, there is no fixed schedule to supply water. Therefore people have to depend on other sources of water to meet their drinking water, cooking, etc. needs. The mismanaged, unplanned and abused use of groundwater has led to groundwater resources overexploited and contaminated with natural and anthropogenic activities. The problem of high dissolved solids, fluoride, metals, cations and anions concentration in groundwater has now become the one of most important, current and alarming toxicological and geo-environmental issues in India and Punjab.

The present study deals with the drinking water qualities in rural areas of district Patiala in the context, pH, electrical conductivity, total dissolved solids, total alkalinity, total suspended solids, total hardness, calcium, magnesium, fluoride, chloride, nitrate and total coliforms. A total of two hundred water samples from villages of district Patiala were analyzed for water quality parameters.

BIS suggested limits for pH are 6.5 to 8.5. WHO (2006) does not specify any health based guideline value for pH of water, although it indicates that in a typical distribution system, the normal range will vary from 6.5 to 8 depending on the composition of water and material used in the system. Drinking water below 6.5 will be more acidic and above 8.5 will be more alkaline. Below and above this limit, consumption of water may lead to health problems like irritation of mucous membrane, bitter taste, etc. Most of the natural waters are usually slightly basic because of the presence of bicarbonates and carbonates of alkali and alkaline earth metals (APHA, 1995). The pH varies according to the composition of water and nature of the construction material used in the distribution system. Extreme pH values can result from accidental spills, treatment breakdowns, and insignificantly cured cement mortar pipe linings. The pH is a critical factor in determining the nature of interactions with some of the transition and heavy metal ions. The pH of water can affects the toxicity of various compounds by changing the ionization equilibrium (Framan, 1981). The pH for the district as a whole is found to be 7.93 with standard deviation 0.2178. The pH of Bhunerheri block lies in between 7.72 to 8.26, Ghanaur block in between 7.69 to 8.54, Nabha block in between 7.75 to 8.54, Rajpura block from

6.98 to 8.17, Patiala block from 6.95 to 8.42, Patran block from 7.86 to 8.36, Samana block from 7.14 to 7.89 and pH of Sanour block lies in between 7.99 to 8.34. The values obtained indicate the very slightly acidic to alkaline nature of water. The pH plays an important role in determining the corrosovity of the water. The pH is inversely proportional to corrosovity. The lower pH increases corrosovity. It has been observed that in some cases increase in pH is accompanied by increase in alkalinity due to increase in bicarbonate, carbonate and hydroxyl ions. Decrease in pH can be caused by increase in the amount of organic carbon, total carbonate and bicarbonate by the use of sewage (Petruzzelli *et al*; 1989; Sotomayor; 1979). In the studies of Cluster-1, pH found to be positively correlated with Electrical Conductivity, Total Dissolved Solids, Total Alkalinity, Magnesium, Nitrate and Total coliforms and negatively correlated with Total Suspended Solids, Total Hardness, Calcium, Fluoride and Chloride. In the studies of Cluster-2, pH found to positively correlated with Total Alkalinity, Magnesium, Fluoride, and Chloride and negatively correlated with Electrical Conductivity, Total Dissolved solids, Total Suspended Solids, Total Hardness, Calcium, Nitrate and Total coliforms. In the studies of Patiala as a whole, pH found to be positively correlated Electrical Conductivity, Total Dissolved Solids, Total Alkalinity, Total Suspended Solids, Magnesium and Fluoride and negatively correlated with Total Hardness, Calcium, Chloride, Nitrate and total Coliforms.

Electrical conductivity for the district Patiala was obtained as 1.345 with standard deviation 0.1631. Mean value of Electrical conductivity of Bhunerheri block was found to be in between 1.25 to 2.44, Ghanaur block was from 0.73 to 3.70, Nabha block from 0.69 to 2.24, Rajpura block from 0.59 to 2.31, Patiala block from 0.74 to 2.57, Patiala block 0.75 to 2.46, Samana block from 0.68 to 1.97 and Electrical Conductivity value of Sanour block was found to be in between 0.75 to 2.08. It is well known that Conductance of water increases with the salts dissolved in it. Total dissolved solids and conductivity can be delineating each other. Conductivity is directly proportional to the dissolved solids. The increase in conductivity may be due to leachate infiltration from refuse dumps (Carlsson, 1997). The concentration of TDS from natural sources have been found to vary from less than 30 mg/L to as much as 6000 mg/L (Trivedi and Goel, 1986). The presence of dissolved solids in water affects its taste. In the water samples of Cluster-1, Electrical Conductivity found to be positively correlated with pH, Total Dissolved Solids, Total alkalinity, Total Suspended Solids, Total Hardness, Calcium, Fluoride, Chloride, Nitrate and Total Coliforms and negatively correlated with Magnesium. In the Cluster-2, Electrical conductivity found to be positively correlated with Total Dissolved Solids, Total alkalinity, Total Suspended Solids, Total hardness, Calcium, Fluoride and negatively correlated with pH, Magnesium, Chloride, Nitrate and Total Coliforms. In the water samples of district Patiala as a whole, pH, Electrical Conductivity found to positively correlated with Total Dissolved

Solids, Total alkalinity, Total Suspended Solids, Total Suspended Solids, Total hardness, Calcium, Fluoride, Chloride and Total Coliforms and negatively correlated with magnesium and Nitrate.

Total Alkalinity is a measure of the ability of the water to neutralize acids. The alkalinity of natural waters is due to the salts of carbonates, bicarbonates, borates, silicates and phosphates along with the hydroxyl ions in the Free State. However the major portion of the alkalinity in natural waters is caused by hydroxide, carbonate and bicarbonate, which may be ranked in order of their association with high pH values. Alkalinity values provide guidance in applying proper doses of chemicals in water and wastewater treatment processes, particularly in coagulation, softening and operational control of anaerobic digestion. Total Alkalinity for the district Patiala was obtained as 322 with standard deviation 52.3967. Mean value of Total Alkalinity of the Bhunerheri block found to be in between 200 to 391, Ghanaur block from 263 to 496, Nabha block from 218 to 492, Rajpura from 139 to 579, Patiala block from 195 to 571, Patran block from 192 to 623, Samana block from 169 to 469 and Total Alkalinity value for the Sanour block found to be 197 to 543. High value of Total Alkalinity in water reservoirs indicates pollution of organic and inorganic nature (Moyle, 1945; Phillip, 1960). Water samples of Cluster-1 show positive correlation of Total Alkalinity with pH, Electrical Conductivity, Total Dissolved Solids, Total Suspended Solids, Total Hardness, Calcium, Magnesium, Nitrate and Total Coliforms and negative correlation with Fluoride and Chloride. Total Alkalinity of Cluster-2 is positively correlated with pH, Electrical Conductivity, Total Dissolved Solids, Total Suspended Solids, Total Hardness, Calcium, Magnesium and fluoride and negatively correlated with Chloride, Nitrate and Total Coliforms. The district Patiala water samples as a whole show positive correlation of Total Alkalinity with pH, Electrical Conductivity, Total Dissolved Solids, Total Suspended Solids, Total Hardness, Calcium, Magnesium and Nitrate and negative correlation with Fluoride, Chloride and Total Coliforms.

Total Dissolved Solids for the district Patiala was obtained as 867 with standard deviation 108.8219. Mean value of Total Dissolved Solids of the Bhunerheri block found to be in between 801 to 1563, Ghanaur block from 467 to 2367, Nabha block from 439 to 1431, Rajpura from 376 to 1476, Patiala block from 473 to 1643, Patran block from 482 to 1572, Samana block from 435 to 1263 and Total Dissolved Solids value for the Sanour block found to be 483 to 1329. Water samples of Cluster-1 show positive correlation of Total Dissolved Solids with pH, Electrical Conductivity, Total Alkalinity, Total Suspended Solids, Total Hardness, Calcium, Fluoride, Chloride, Nitrate and Total Coliforms and negative correlation with Magnesium. Total Dissolved Solids of Cluster-2 is positively correlated with Electrical Conductivity, Total Suspended Solids, Total Hardness, Calcium and fluoride and negatively correlated with pH, Magnesium, Chloride, Nitrate and Total Coliforms. The district Patiala water samples as a whole show positive correlation of Total Dissolved Solids with pH, Electrical Conductivity, Total Alkalinity, Total Suspended Solids, Total Hardness, Calcium, Fluoride, Chloride and Total Coliforms and negative correlation with Magnesium and Nitrate.

Total Suspended Solids for the district Patiala was obtained as 0.642 with standard deviation 0.0859. Mean value of Total Suspended Solids of the Bhunerheri block found to be in between 0.559 to 0.589, Ghanaur block from 0.317 to 1.608, Nabha block from 0.343 to 1.118, Rajpura from 0.294 to 1.153, Patiala block from 0.370 to 1.284, Patran block from 0.377 to 1.228, Samana block from 0.340 to 0.987 and Total Suspended Solids value for the Sanour block found to be 0.377 to 1.038. Water samples of Cluster-1 show positive correlation of Total Suspended Solids with Electrical Conductivity, Total Dissolved Solids, Total Alkalinity, Total Hardness, Calcium, Magnesium, Chloride, Nitrate and Total Coliforms and negative correlation with pH, Fluoride. Total Suspended Solids of Cluster-2 is positively correlated with Electrical Conductivity, Total Dissolved Solids, Total Alkalinity, Total Hardness, Calcium and fluoride and negatively correlated with pH, Magnesium, Chloride, Nitrate and Total Coliforms. The district Patiala water samples as a whole show positive correlation of Total Suspended Solids with pH, Electrical Conductivity, Total Dissolved Solids, Total Alkalinity, Total Hardness, Calcium, Magnesium, Nitrate and Total Coliforms and negative correlation with Fluoride and Chloride.

Chloride in surface and groundwater originates from both natural and anthropogenic sources. The presence of chloride in natural waters can be attributed to dissolution of salt deposits, discharges of effluents from chemical industries, oil well operations and seawater intrusion in coastal areas. Each of these sources may result in local contamination of both surface water and groundwater. A high content of chloride gives salty taste to water. The salty taste produced by chloride depends on the chemical composition of the water. A concentration of 250 mg/L may be detectable in some waters containing sodium ions. On the other hand, the typical salty taste may be absent in water containing 1000 mg/L chloride when calcium and magnesium ions are predominant. High chloride content may harm metallic pipes and structures as well as agricultural plants. People who are not accustomed to high chlorides in water are subjected to laxative effects (Prakash and Rao 1989). The acceptable desirable limit of chloride is 250 mg/L. Beyond this limit, taste, corrosion and palatability are affected. The mean value obtained for Chloride contents for district Patiala is 89.04 with standard deviation 27.3507. Mean value of Chloride contents of the Bhunerheri block found to be in between 86 to 117, Ghanaur block from 39 to 175, Nabha block from 53 to 176, Rajpura from 56 to 232, Patiala block from 78 to 228, Patran block from 31 to 106, Samana block from 30 to 91 and

Chloride contents value for the Sanour block found to be 42 to 117. In the water samples of Cluster-1, Chloride found to be positively correlated with Electrical Conductivity, Total Dissolved Solids, Total Suspended Solids, Total Hardness, Calcium and Fluoride and negatively correlated with pH, Total Alkalinity, Magnesium, Nitrate and Total coliforms. In the Cluster-2, Chloride found to be positively correlated with pH, Nitrate and Total Coliforms and negatively correlated with Electrical Conductivity, Total Dissolved Solids, Total Alkalinity, Total Suspended Solids, Total Hardness, calcium, Magnesium and Fluoride. In the water samples of district Patiala as a whole, Chloride found to positively correlated with Electrical Conductivity, Total Dissolved Solids, Total hardness, Calcium, Nitrate and Total Coliforms and negatively correlated with pH, Total Alkalinity, Total Suspended Solids, Magnesium and Fluoride.

Water hardness is a traditional measure of the capacity of water to precipitate soap. Hardness of water is not a specific constituent but is a variable and complex mixture of cations and anions. It is caused by dissolved polyvalent metallic ions. In fresh water, the principal hardness-causing ions are calcium and magnesium which precipitate soap. Other polyvalent cations also may precipitate soap, but often are in complex forms, frequently with organic constituents, and their role in water hardness may be minimal and difficult to define. Total hardness is defined as the sum of the calcium and magnesium concentration, both expressed as $CaCO_3$, in mg/L. The higher concentration of Total Hardness in water samples may be due to dissolution of bivalent cations from sedimentary rocks, seepage and run off from soil. The mean value obtained for district Patiala is 371 with standard deviation 78.3345. Mean value of Total Hardness of the Bhunerheri block found to be in between 207 to 336, Ghanaur block from 159 to 504, Nabha block from 220 to 516, Rajpura from 188 to 738, Patiala block from 237 to 822, Patran block from 241 to 786, Samana block from 218 to 632 and Total Hardness value for the Sanour block found to be 193 to 532. In the water samples of Cluster-1, Total Hardness found to be positively correlated with electrical conductivity, total dissolved solids, total alkalinity, total suspended solids, calcium, chloride, nitrate and total coliforms and negatively correlated with pH, magnesium and fluoride. In the Cluster-2, Total Hardness found to be positively correlated with electrical conductivity, total dissolved solids, total alkalinity, total suspended solids, calcium, and negatively correlated with pH, magnesium, fluoride, chloride, nitrate and total coliforms. In the water samples of district Patiala as a whole, Total Hardness found to positively correlated with electrical conductivity, total dissolved solids, total alkalinity, total suspended solids, calcium, chloride, nitrate and total coliforms and negatively correlated with pH, magnesium, fluoride. The strong correlation-ship between these parameters could be due to changes in land use namely deforestation, disruption in internal sources of hardness and alkalinity, climatic factor and industrialization (Kumar, 1993).

The mean value obtained for Calcium ions for district Patiala is 79.3 with standard deviation 39.4664. Mean value of Calcium ions of the Bhunerheri block found to be in between 30 to 65, Ghanaur block from 16 to 74, Nabha block from 22 to 72, Rajpura from 50 to 197, Patiala block from 63 to 219, Patran block from 64 to 210, Samana block from 58 to 168 and Calcium ions value for the Sanour block found to be 35 to 95. In the water samples of Cluster-1, Calcium ions found to be positively correlated with electrical conductivity, total dissolved solids, total alkalinity, total suspended solids, total hardness, fluoride, chloride, nitrate and total coliforms and negatively correlated with pH, magnesium. In the water samples of Cluster-2, Calcium ions found to be positively correlated with total hardness and total coliforms and negatively correlated with pH, electrical conductivity, total dissolved solids, total alkalinity, total suspended solids, magnesium, fluoride, chloride, nitrate. In the water samples of district Patiala as a whole, Calcium ions found to be positively correlated with electrical conductivity, total dissolved solids, total alkalinity, total suspended solids, total hardness, chloride, nitrate and total coliforms and negatively correlated with pH, magnesium, fluoride.

The mean value obtained for Magnesium ions for district Patiala is 41.94 with standard deviation 11.0318. Mean value of Magnesium ions of the Bhunerheri block found to be in between 22 to 63, Ghanaur block from 29 to 85, Nabha block from 40 to 89, Rajpura from 15 to 60, Patiala block from 19 to 67, Patran block from 20 to 64, Samana block from 18 to 51 and Magnesium ions value for the Sanour block found to be 26 to 72. In the water samples of Cluster-1, Magnesium ions found to be positively correlated with pH, total alkalinity, total suspended solids, and negatively correlated with electrical conductivity, total dissolved solids, total hardness, calcium, magnesium, fluoride, chloride, nitrate and total coliforms. In the water samples of Cluster-2, Magnesium ions found to be positively correlated with pH, total alkalinity and fluoride and negatively correlated with electrical conductivity, total dissolved solids, total suspended solids, total hardness, calcium, chloride, nitrate and total coliforms. In the water samples of district Patiala as a whole, Magnesium ions found to be positively correlated with pH, total alkalinity, total suspended solids and fluoride and negatively correlated with electrical conductivity, total dissolved solids, total hardness, calcium, chloride, nitrate and total coliforms.

During the last four decades, the high fluoride concentration in water resources and the resultant disease "Fluorosis" is being highlighted considerably throughout the world. According to the "Survey of Status of Drinking Water in Rural Habitation" conducted by Rajiv Gandhi National Drinking Water Mission in 1993, there are 9741 villages and 6819 habitations having fluoride content more than 1.5 mg/L in groundwater resources. The desirable limit of fluoride is 1.0 mg/L, and permissible up to 1.5 mg/L in case of non-availability of other water sources. The main sources of drinking water in study area are hand pumps, tube well and dug-wells. In general, it has been observed that fluoride present in the groundwater may be dissolved from geologic conditions while surface water usually contains lesser fluoride content except when contaminated by industrial water. Thus the fluoride accumulation in groundwater in different areas varies according to source of water (surface or subterranean), the geological formation of the area, amount of rainfall and quantity of water lost by evaporation. The various factors that govern the release of fluoride in natural water by fluoride bearing minerals and rocks are the basic chemical composition of water, presence and accessibility of fluoride and the time of contact between the source of mineral and water. In the present study no water sample showed fluoride content higher than the permissible limit. According to WHO (1997), permissible limit for fluoride in drinking water is 1.0 mg/L, whereas, USPHS (1962) has set a range of allowable concentrations for fluoride in drinking water for a region depending on its climatic conditions because the amount of water consumed and consequently the amount of fluoride ingested being influenced by the air temperature. Lesan (1987) suggests a limit of fluoride in drinking water as low as 0.6 mg/L under tropical conditions. The mean value obtained for Fluoride for district Patiala is 0.338 with standard deviation 0.1065. Mean value of Fluoride of the Bhunerheri block found to be in between 0.199 to 1.030, Ghanaur block from 0.212 to 0.563, Nabha block from 0.293 to 0.655, Rajpura from 0.076 to 0.299, Patiala block from 0.192 to 0.665, Patran block from 0.195 to 0.637, Samana block from 0.176 to 0.512 and Fluoride value for the Sanour block found to be 0.130 to 0.358. In the water samples of Cluster-1, Fluoride found to be positively correlated with electrical conductivity, total dissolved solids, calcium, chloride, nitrate and total coliforms and negatively correlated with pH, total alkalinity, total suspended solids, total hardness, magnesium. In the water samples of Cluster-2, Fluoride found to be positively correlated with pH, electrical conductivity, total dissolved solids, total suspended solids, magnesium and negatively correlated with total hardness, calcium, chloride, nitrate and total coliforms. In the water samples of the district Patiala as a whole, Fluoride found to be positively correlated with pH, electrical conductivity, total dissolved solids and magnesium and negatively correlated with total alkalinity, total hardness, chloride, nitrate and total coliforms.

The toxicity of nitrates to humans is thought to be solely the consequence of its reduction to nitrite (Gulis et al., 2002). The results of investigations of Yang et al. (1998) showed that there is a significant positive association between drinking water nitrate exposure and gastric cancer mortality. Nitrate is used mainly in inorganic fertilizers (Landon et al., 2000; WHO, 1996). The value of nitrates may be high as a result of agricultural runoff, refuge dump runoff, or contamination with human or animal wastes (Neal et al., 2000). The mean value obtained for Nitrate for district Patiala 3.376 with standard deviation 1.8999. Mean value of Nitrate of the Bhunerheri block found to be in between 2.013 to 3.927, Ghanaur block from 1.173 to 5.947, Nabha block from 1.103 to 3.595, Rajpura from 3.837 to 15.061, Patiala block from 1.587 to 5.513, Patran block from 2.434 to 7.939, Samana block from 1.093 to 3.173 and Nitrate value for the Sanour block found to be 1.621 to 4.460. In the water samples of Cluster-1, Nitrate found to be positively correlated with pH, electrical conductivity, total dissolved solids, total alkalinity, total suspended solids, total hardness, calcium, fluoride, and total coliforms and negatively correlated with magnesium and chloride. In the water samples of Cluster-2, Nitrate found to be positively correlated with calcium, chloride and total coliforms and negatively correlated with pH, electrical conductivity, total dissolved solids, total alkalinity, total suspended solids, total hardness, magnesium, fluoride. In the water samples of the district Patiala as a whole, Nitrate found to be positively correlated with total alkalinity, total suspended solids, total hardness, calcium, chloride and total coliforms and negatively correlated with pH, electrical conductivity, total dissolved solids, magnesium, and fluoride.

Coliforms are bacteria that are always present in the digestive tracts of animals, including humans, and are found in their wastes. They are also found in plant and soil material. The most basic test for bacterial contamination of a water supply is the test for total coliform bacteria. Total coliform counts give a general indication of the sanitary condition of a water supply. Immediate investigative action must be taken if total coliforms are detected. It is recognized that the great majority of rural water supplies, especially in developing countries, faecal contamination is wide spread (WHO, 2006). BIS suggests that drinking water should be free from pathogens to avoid water borne diseases like diarrhea, cholera, hepatitis etc. WHO (2004) suggests that it may be useful to classify drinking water systems into categories that are predefined depending on the risks associated with the drinking water, the order of priorities placed, and the local circumstance. Water is classified into five types as per needs. Class-A, B and C type of water can be used for drinking purposes and these types may contain up to 50, 500, and 5000 Total Coliforms. But water is to be consumed after disinfection and treatment if contains pathogens to avoid any type of water borne diseases. The mean value obtained for Total Coliforms for district Patiala 98 with standard deviation 49.5824. Mean value of Total Coliforms of the Bhunerheri block found to be in between 81 to 157, Ghanaur block from 47 to 238, Nabha block from 22 to 72, Rajpura from 102 to 402, Patiala block from 42 to 147, Patran block from 65 to 212, Samana block from 29 to 85 and Total Coliforms value for the Sanour block found to be 50 to 137. In the water samples of Cluster-1, Total Coliforms found to be positively correlated with pH, electrical conductivity, total dissolved solids, total alkalinity, total suspended solids, total hardness, calcium, fluoride, chloride and nitrate and negatively correlated with magnesium. In the samples of Cluster-2, Total coliforms found to be positively correlated with calcium and nitrate and negatively correlated with pH, electrical conductivity, total dissolved solids, total alkalinity total suspended solids, total hardness, magnesium, and fluoride. In the water samples of the district Patiala as a whole, Total Coliforms found to be positively correlated with electrical conductivity, total dissolved solids, total suspended solids, total hardness, calcium, chloride and nitrate and negatively correlated with pH, total alkalinity, magnesium and fluoride.

The quality of drinking water is a complex issue, but is vital for public health. Poor water quality is responsible for an estimated death of five million children annually (Holgate, 2000; Thurman et al., 1998). The present situation of water quality management is far from satisfactory (Huang and Xia, 2001). Perhaps, public and decision makers of most of the developing countries of the world are not aware of the gravity of the situation. The vast majority of diarrheal disease in the world (88%) is attributable to unsafe water, sanitation and poor hygiene actions (Nath *et al.*, 2006). Earthenware vessels showed significantly higher levels of contamination (Vanderslice and Briscoe, 1993). A community should be empowered with alternate means to treat drinking water in order to meet the challenges of safe drinking water for every home. Willmitzer (2000) suggests that water protection is always cheaper than expensive water body restoration and water treatment; therefore priority should be given to the avoidance of contaminants at the point of origin.

According to Au et al. (2000), public control and management of the environment and official decisions about environmental issues increasingly require public environmental monitoring by citizens, because official monitoring aims to determine if regulations are being broken, thus relying on accurate measurements obtained by the court acceptable procedures while public monitoring aims to determine if a site of concern to local citizens is a source of pollution, which can be obtained by a variety of scientifically sound and reliable methods. Edberg et al. (1997) also stressed on complete monitoring of drinking water comprising of the microbiological as well as physicochemical monitoring to establish a long term history of the source.

Chapter-7

CONCLUSION AND RECOMMENDATIONS

7.1 CONCLUSION

The environmental contamination of drinking water at source, through the conveyance system and even at the users' level, can spread infectious diseases like cholera, typhoid, hepatitis, dysentery, worms etc. Also, harmful materials such as heavy metals, pesticides etc. can reach drinking water by various routes and water quality may deteriorate (WHO, 1995). In India, and also in Punjab, due to rapid population growth, unsanitary disposal of wastes and other human activities, most of the water sources are becoming polluted. The prevailing practices of open defecation, unscientific disposal of human wastes and agricultural practices in most of the rural villages have increased the level of microbiological contamination in the water from streams, springs and ground sources. Water quality problems caused by physical and chemical parameters have huge impacts on public health when the concentrations are high. In many regions, the drawdown of the water table every year is another serious problem, with some shallow tube wells becoming non-functional. Many of the major problems associated with tube-well systems and gravity flow systems can be attributed to lack of feasibility studies and poor design, which does not take water quality into consideration. Responsibility, commitment, guidelines and actions are not clearly defined within policies, legislations or institutional frame works at either the National or the Grass Roots Level. Water management has always been a contentious and tricky affair in India due to socio-economic-political and ecological reasons. Factors like caste-class differences, heterogeneity of farmers, rural–urban dichotomy, and extreme different ecological conditions have influenced the water management. To complicate further, vote bank politics, lack of coordination between irrigation bureaucracy, policy making and various sectoral departments carrying out their own water programmes, have affected water management in a diverse manner to people. Understanding how these different policies and programs influence water management at the community level is one of the unexplored issues.

For the provision of quality water supplies, the concerned authorities have to give special consideration to the preparation of drinking water quality guidelines and inventories for both quality and quantity of water coverage. The quality of water that is potable should be properly defined. Treatment of water in all rural supply systems to improve the water quality seems not to be economically feasible or manageable at the national level. This really needs intensive, detailed, investigation and research to determine appropriate and sustainable technological options, which take into consideration the socio-economic and traditional values of the country. Thus, water quality assessment of specific quality parameters is essential at the various stages of development. This in turn will help in the development of appropriate strategic planning and remedial action for water quality improvements. There is a clear need for an effective water quality monitoring and surveillance program to ensure a safe and sustainable water supply system.

This research was carried out with the purpose to document and analyze the physicochemical and microbiological quality assessment of drinking water in rural areas of district Patiala. The investigation on survey and characterization on drinking water qualities of villages of the eight blocks of district Patiala viz. Bhunerheri, Ghanaur, Nabha, Patiala, Patran, Rajpura, Samana and Sanour, was carried for water quality parameters like temperature, pH, electrical conductivity, total dissolved solids, total suspended solids, total alkalinity, total hardness, calcium, magnesium, chloride, fluoride, nitrate and total coliforms. A total of two hundred samples were collected from hand pumps and tube-wells, keeping in mind the habits and sources of drinking water of the people residing in the villages.

Thirty water samples were studied of the Bhunerheri block (Table No.5.2 and Table No.5.10). All the samples of the block were found within the limits suggested by BIS as per pH (6.5-8.5) is concerned. Water of the block is alkaline. None of the samples was above the BIS permissible limit (2000) for total dissolved solids. However, one sample was found above the WHO limit (1500). All samples showed total alkalinity above the desirable limit (200) of BIS but below permissible limit (600). Only five samples were above the desirable limit (300) as suggested by BIS but below permissible limit (600) for total hardness, while all the samples were above the WHO limit (100). Calcium contents are below the desirable limit suggested by BIS and WHO (75) while fourteen samples had magnesium contents above the desirable limit (30) of BIS but below the permissible limit, while only two samples were found above the WHO limit (50). Fluoride in all the samples except one was below the desirable limit (1.0) suggested by BIS and WHO. Chloride contents in all the samples were below desirable limit (250) as suggested by BIS and WHO. Nitrate was also below the desirable limit (45) suggested by BIS. As per total coliforms are concerned, drinking water of the Bhunerheri block can be classified as type-B water which needs proper disinfection and treatment before drinking.

Thirty water samples from the Ghanaur block were studied Table No.5.3 and Table No.5.11). All the samples of the block were found within the limits suggested by BIS as per pH (6.5-8.5) is concerned. Water of the block is alkaline. One sample was found above the BIS permissible limit (2000) for total dissolved solids. However, two samples found above the WHO limit (1500). All samples showed total alkalinity above the desirable limit (200) of BIS but below permissible limit (600). Eleven samples were above the desirable limit (300) as suggested by BIS but below permissible limit (600) for total hardness, while all the samples were above the WHO limit (100). Calcium contents are below the desirable limit suggested by BIS and WHO (75) while four samples had magnesium contents below the desirable limit (30) of BIS and four samples had magnesium above the permissible limit (75), while thirteen samples were found above the WHO limit (50). Fluoride in all samples was below the desirable limit (1.0) suggested by BIS and WHO. Chloride contents in all the samples were below desirable limit (250) as suggested by BIS and WHO. Nitrate was also below the desirable limit (45) suggested by BIS. As per total coliforms are concerned, drinking water of the Ghanaur block can be classified as type-B water which needs proper disinfection and treatment before drinking.

Thirty water samples from the Nabha block were studied (Table No.5.4 and Table No.5.12). All the samples of the block were found within the limits suggested by BIS as per pH (6.5-8.5) is concerned. Water of the block is alkaline. None of the samples was found above the BIS permissible limit (2000) and WHO limit (1500) for total dissolved solids. All samples showed total alkalinity above the desirable limit (200) of BIS but below permissible limit (600). Thirteen samples were above the desirable limit (300) as suggested by BIS but below permissible limit (600) for total hardness, while all the samples were above the WHO limit (100). Calcium contents are below the desirable limit suggested by BIS and WHO (75) while all samples had magnesium contents above the desirable limit (30) of BIS and ten samples had magnesium above the permissible limit (75), while fifteen samples were found above the WHO limit (50). Fluoride in all samples was below the desirable limit (1.0) suggested by BIS and WHO. Chloride contents in all the samples were below desirable limit (250) as suggested by BIS and WHO. Nitrate was also below the desirable limit (45) suggested by BIS. As per total coliforms are concerned, drinking water of the Nabha block can be classified as type-A (twenty five samples having total coliforms below 50) and type-B (five samples with total coliforms above 50) water which needs proper disinfection and treatment before drinking.

Thirty water samples from the Rajpura block were studied (Table No.5.5 and Table No.5.13). All the samples of the block were found within the limits suggested by BIS as per pH (6.5-8.5) is concerned. Water of the block is slightly alkaline except one sample (pH, 6.98). No sample was found above the BIS permissible limit (2000) and WHO limit (1500) for total dissolved solids. Nineteen samples showed total alkalinity above the desirable limit (200) of BIS and eleven samples showed above desirable limit but below permissible limit (600). Seventeen samples were above the desirable limit (300) as suggested by BIS and three samples were above the permissible limit (600) for total hardness, while all the samples were above the WHO limit (100). Calcium contents of twenty samples were found above the desirable limit suggested by BIS and WHO (75) but below the permissible limit (200). While fourteen samples had magnesium contents above the desirable limit (30) of BIS and none had magnesium above the permissible limit (75), while two samples were found above the WHO limit (50). Fluoride in all samples was below the desirable limit (1.0) suggested by BIS and WHO. Chloride contents in all the samples were below desirable limit (250) as suggested by BIS and WHO. Nitrate was also below the desirable limit (45) suggested by BIS. As per total coliforms are concerned, drinking water of the Rajpura block can be classified as type-B water which needs proper disinfection and treatment before drinking.

Twenty water samples from the Patiala block were studied (Table No.5.6 and Table No.5.14). All the samples of the block were found within the limits suggested by BIS as per pH (6.5-8.5) is concerned. Water of the block is alkaline except for two samples with pH less than seven. No sample was found above the BIS permissible limit (2000) for total dissolved solids. However, one samples found above the WHO limit (1500). Eighteen samples showed total alkalinity above the desirable limit (200) of BIS but below permissible limit (600). Sixteen samples were above the desirable limit (300) as suggested by BIS and four samples were above the permissible limit (600) for total hardness, while all the samples were above the WHO limit (100). Calcium contents of four samples were below the desirable limit suggested by BIS and WHO (75) and one sample was found above the permissible limit (200). While seven samples had magnesium contents below the desirable limit (30) of BIS and none of the samples had magnesium above the permissible limit (75), while four samples were found above the WHO limit (50). Fluoride in all samples was below the desirable limit (1.0) suggested by BIS and WHO. Chloride contents in all the samples were below desirable limit (250) as suggested by BIS and WHO. Nitrate was also below the desirable limit (45) suggested by BIS. As per total coliforms are concerned, drinking water of the Patiala block can be classified as type-A (four samples have total coliforms below 50) and type-B (sixteen samples with total coliforms above 50) water which needs proper disinfection and treatment before drinking.

Twenty water samples from the Patran block were studied (Table No.5.7 and Table No.5.15). All the samples of the block were found within the limits suggested by BIS as per pH (6.5-8.5) is concerned. Water of the block is alkaline. No sample was found above the BIS permissible limit (2000) for total dissolved solids. However, one samples found above the WHO limit (1500). Eighteen samples showed total alkalinity above the desirable limit (200) of BIS and two above the permissible limit (600). Fifteen samples were above the desirable limit (300) as suggested by BIS and five samples were above the permissible limit (600) for total hardness, while all the samples were above the WHO limit (100). Calcium contents of three samples were found below the desirable limit suggested by BIS and WHO (75) and one sample was found with calcium contents above the permissible limit (200). While seven samples had magnesium contents below the desirable limit (30) of BIS and no samples had magnesium above the permissible limit (75), while five samples were found above the WHO limit (50). Fluoride in all samples was below the desirable limit (1.0) suggested by BIS and WHO. Chloride contents in all the samples were below desirable limit (250) as suggested by BIS and WHO. Nitrate was also below the desirable limit (45) suggested by BIS. As per total coliforms are concerned, drinking water of the Patran block can be classified as type-B water which needs proper disinfection and treatment before drinking.

Twenty water samples from the Samana block were studied (Table No.5.8 and Table No.5.16). All the samples of the block were found within the limits suggested by BIS as per pH (6.5-8.5) is concerned. Water of the block is alkaline. No sample was found above the BIS permissible limit (2000) and WHO limit (1500) for total dissolved solids. Two samples showed total alkalinity below the desirable limit (200) of BIS and none above the permissible limit (600). Six samples were below the desirable limit (300) as suggested by BIS and one sample was found above the permissible limit (600) for total hardness, while all the samples were above the WHO limit (100). Calcium contents of three samples were found below the desirable limit suggested by BIS and WHO (75) and no sample was found with calcium contents above the permissible limit (200). While seven samples had magnesium contents below the desirable limit (30) of BIS and no samples had magnesium above the permissible limit (75), while one sample was found above the WHO limit (50). Fluoride in all samples was below the desirable limit (1.0) suggested by BIS and WHO. Chloride contents in all the samples were below desirable limit (250) as suggested by BIS and WHO. Nitrate was also below the desirable limit (45) suggested by BIS. As per total coliforms are concerned, drinking water of the Samana block can be classified as type-A (ten samples with total coliforms below 50) and type-B (ten samples with total coliforms above 50) water which needs proper disinfection and treatment before drinking.

Twenty water samples from the Sanour block were studied (Table No.5.9 and Table No.5.17). All the samples of the block were found within the limits suggested by BIS as per pH (6.5-8.5) is concerned. Water of the block is alkaline. No sample was found above the BIS permissible limit (2000) and WHO limit (1500) for total dissolved solids. All samples except one showed total alkalinity above the desirable limit (200) of BIS and none above the permissible limit (600). Sixteen samples were above the desirable limit (300) as suggested by BIS and no sample was above the permissible limit (600) for total hardness, while all the samples were above the WHO limit (100). Calcium contents of thirteen samples were found below the desirable limit suggested by BIS and WHO (75) and none sample was found with calcium contents above the permissible limit (200). While one samples had magnesium contents below the desirable limit (30) of BIS and no samples had magnesium above the permissible limit (75), while thirteen samples were found above the WHO limit (50). Fluoride in all samples was below the desirable limit (1.0) suggested by BIS and WHO. Chloride contents in all the samples were below desirable limit (250) as suggested by BIS and WHO. Nitrate was also below the desirable limit (45) suggested by BIS. As per total coliforms are concerned, drinking water of the Sanour block can be classified as type-B water which needs proper disinfection and treatment before drinking.

Two quality parameters, Total Dissolved Solids (Table No.5.19 and Figure No.5.30 and Total Hardness (Table No.5.18 and Figure No.5.2) are of general importance. On the basis of physicochemical monitoring of the samples studied, it may be concluded that the water of all the samples studied is of hard type and none of the samples showed soft or moderately hard type water quality. Water with Total Hardness below 75 can be considered soft, 75 to 150 moderately hard, 151 to 300 hard and water with Total Hardness above 300 is termed as very hard (Matalas and Reiher, 1967 & Sawyer and McCarthy, 1967). Table No. 5.18 and Figure No. 5.2 indicates that 80% is hard water and 20% of water is very hard in the Bhunerheri block. Ghanaur block has 67% hard water and 33% very hard water. Water of Nabha block can be classified as 50% hard and 50% very hard. Rajpura block has 44% hard and 56% very hard water. Patiala block has 25% hard and 75% very hard water. Patran block contains 25% hard and 75% very hard water. Water of Samana block can be classified as 20% hard and 80% very hard. Sanour block contains 10% hard water and 90% very hard water. Consumption of very soft and very hard water is not good for human beings. Very soft water does not contain much solids dissolved, needed to digest the food while very hard water contains much solids dissolved in it and can create problems. So the water samples come under Slightly Hard and Moderately Hard category, can be consumed safely, if otherwise found suitable. As per Total Dissolved Solids are concerned, safe and desirable limit as prescribed by BIS and WHO is 500, but water with Total Dissolved Solids up to 1500 (WHO) and 2000 (BIS) can be consumed if no other alternative is available. On the basis of physicochemical monitoring of the samples studied (Table No.5.19 and Figure No.5.3), it may be concluded that none of the samples showed Total Dissolved Solids below 300 means no water can be termed as good. Water with Total Dissolved Solids 300 to 500 can be considered fair, 500 to 900 average, 900 to 1200 poor, 1200 to 2000 very poor and water with Total Dissolved Solids above 1200 is termed as unacceptable (Matalas and Reiher, 1967). On the basis of Total Dissolved Solids, it was found that none water samples in Bhunerheri block, 17% water samples in Ghanaur block, 44% water samples in Nabha block, 15% water samples in Patran block, 25% water samples in Patiala block, 33% water samples in Rajpura block, 10% water samples in Samana block and 5% water samples in

Sanour block may be termed as fair. Average plus poor category include 80% water samples in Bhunerheri block, 70% water samples in Ghanaur block, 44% water samples in Nabha block, 60% water samples in Patran block, 60% water samples in Patiala block, 57% water samples in Rajpura block, 85% water samples in Samana block and 75% water samples in Sanour block. All the block include some water samples in very poor category and one water samples was found in Ghanaur block in unacceptable category.

7.2 DATA MANIPULATION AND REGIONAL WATER SYSTEM REPORT (RWSR) (Calculating the test data, test ratio and indicator value)

An information system "The WATER DIALOGUE" (Latour et al., 1996) can be used to retrieve the great diversification of data required for integral water management in an uncomplicated manner. The application provides processing facilities and a simple presentation of data relating to the condition and use of water systems. In addition to information regarding the physical, chemical and biological quality of the waters, water dialogue also provides input indicators, the costs involved in the water policy and information with respect to the functional aspects. Within the Regional Water System Report a "regional water system picture" has been introduced as a method to present the results (van der Straten et al., 1998). RWSR follows the structure to define water systems, functions in the water systems, issues/aspects of interest and indicators for these issues to identify the problems. The data manipulation step in RWSR consists of 2 activities: calculating the test value from the measuring data and calculating the indicator value. Basically this is the standard procedure for testing for compliance (Niederlander et al., 1996). In the first step the measured value in a water system is converted to a test value (mean value). The next step is to determine how far the test value deviates from the standard. To achieve this, the test value has been divided by the corresponding standard, resulting in a dimensionless value. After this point the deviation is converted into an individual (water-) index. This is a two-step approach. The deviation is limited (called test ratio) between 1 and 11. A test ratio less than 1 is equated to 1, while values greater than 11 are equated to 11. This test ratio is converted to a water index somewhere in the range between 0 (major deviation from the standard) and 100 (no problem). The formula (water index = 110 - (10*test ratio) shows that this step is a linear transformation. The calculated water index is classified into an indicator score as given in the Table No.7.1.

The overall quality of drinking water as rated by the water quality index and indicator value for Bhunerheri (Table N0.7.2) block is 3 (Satisfy Function Needs Minimal), Ghanaur (Table No.7.4) block as 3 (Satisfy Function Needs Minimal), Nabha (Table No.7.4) block as 4 (Satisfy Function Needs Optimal), Rajpura (Table No.7.5) block as 3 (Satisfy Function Needs Minimal), Patiala (Table No.7.6) block as 3 (Satisfy Function Needs Minimal), Patran (Table

No.7.7) block as 3 (Satisfy Function Needs Minimal), Samana (Table No.7.8) block as 4 (Satisfy Function Needs Optimal), and Sanour (Table No.7.9) block as 3 (Satisfy Function Needs Minimal). The drinking water quality of Patiala district (Table No.7.10) as a whole can be rated as 3 (Satisfy Function Needs Minimal) on the basis of water quality index and indicator value. Figure No.7.1 to 7.9 indicates the factors and their degree which deteriorate the water quality in the different blocks of the district Patiala.

| | **Table 7.1: From Water Index to Indicator Value and Legend (Water Dialogue/ RWSR)** |
Sr. No.	Water Index	Deviation of the Standard	Indicator Value	Legend	Decoding	Signal
1	< 53	> 5.7 ×	1	RED	Deviates Severely From Function Needs	
2	53.01-79.99	3-5.7 ×	2	ORANGE	Does Not Satisfy Function Needs	
3	80.01-92.99	1.7-3 ×	3	YELLOW	Satisfy Function Needs Minimal	
4	93.01-99.99	< 1.7 ×	4	GREEN	Satisfy Function Needs Optimal	
5	100	null	5	BLUE	Satisfy Function Needs	

(Source: Witteveen en Bos, 1999)

Table 7.2: Water Quality Index (Bhunerheri Block)							
Parameter	Test Value	Standard Value	Deviation	Test Ratio	Water Quality Index	Indicator Value	Signal
pH	8	7.5	1.067	1.067	99.33	4	
TDS	987	500	1.974	1.974	90.26	3	
TA	247	200	1.235	1.235	97.65	4	
TH	259	300	0.863	1	100.00	5	
Ca	46.2	75	0.616	1	100.00	5	
Mg	34.7	30	1.157	1.157	98.43	4	
F	0.5	1	0.500	1	100.00	5	
Cl	99.7	250	0.399	1	100.00	5	
NO3-	2.5	45	0.056	1	100.00	5	
TC	99	10	9.900	9.900	11.00	1	
Overall Average					**89.67**	**3**	

Figure 7.1: Ratio of Parameters degrading water quality of Bhunerheri Block

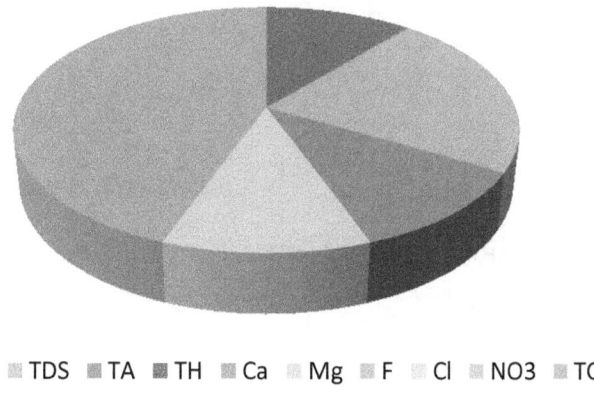

■ pH ■ TDS ■ TA ■ TH ■ Ca ■ Mg ■ F ■ Cl ■ NO3 ■ TC

Table 7.3: Water Quality Index (Ghanaur Block)							
Parameter	Test Value	Standard Value	Deviation	Test Ratio	Water Quality Index	Indicator Value	Signal
pH	8.1	7.5	1.080	1.080	99.20	4	
TDS	853	500	1.706	1.706	92.94	3	
TA	319	200	1.595	1.595	94.05	4	
TH	278	300	0.927	1	100.00	5	
Ca	28.8	75	0.384	1	100.00	5	
Mg	50	30	1.667	1.667	93.33	4	
F	0.3	1	0.300	1	100.00	5	
Cl	69.4	250	0.278	1	100.00	5	
NO3-	2.1	45	0.047	1	100.00	5	
TC	86	10	8.600	8.600	24.00	1	
Overall Average					90.35	3	

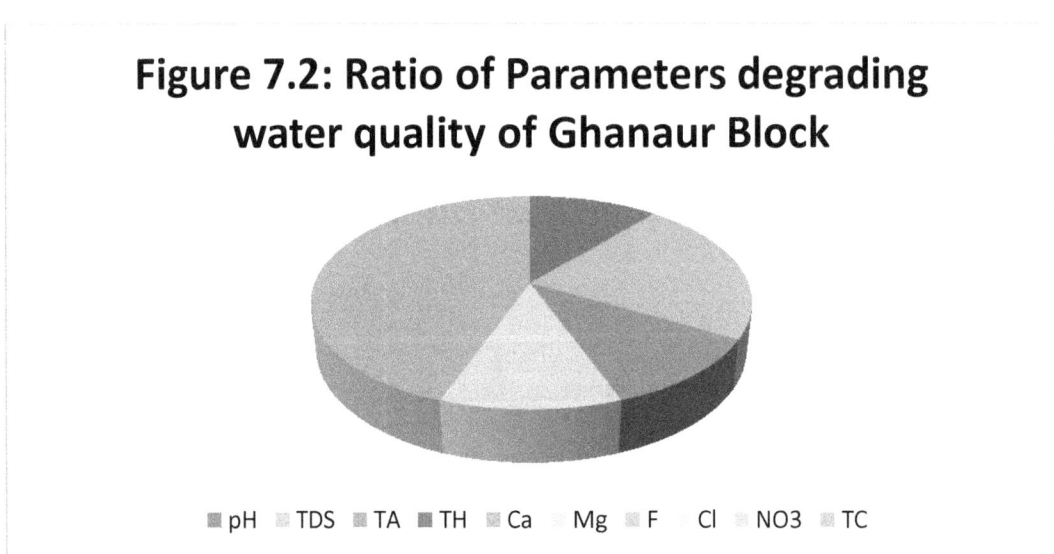

Figure 7.2: Ratio of Parameters degrading water quality of Ghanaur Block

pH TDS TA TH Ca Mg F Cl NO3 TC

Table 7.4: Water Quality Index (Nabha Block)							
Parameter	Test Value	Standard Value	Deviation	Test Ratio	Water Quality Index	Indicator Value	Signal
pH	8.09	7.5	1.079	1.079	99.21	4	
TDS	717	500	1.434	1.434	95.66	4	
TA	293	200	1.465	1.465	95.35	4	
TH	339	300	1.130	1.130	98.70	4	
Ca	35.9	75	0.479	1	100.00	5	
Mg	60.6	30	2.020	2.020	89.80	3	
F	0.413	1	0.413	1	100.00	5	
Cl	86.9	250	0.348	1	100.00	5	
NO3-	1.802	45	0.040	1	100.00	5	
TC	36	10	3.600	3.600	74.00	2	
Overall Average					95.27	4	

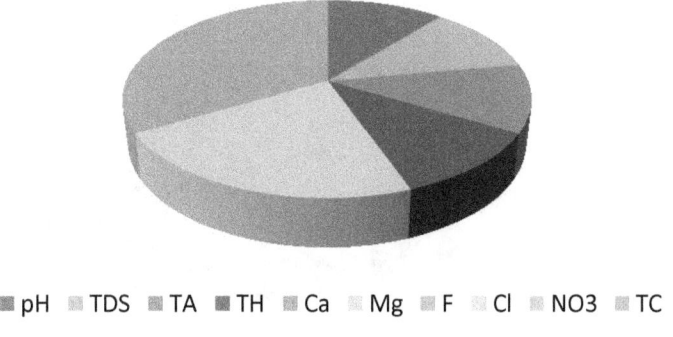

Figure 7.3: Ratio of Parameters degrading water quality of Nabha Block

pH ▪ TDS ▪ TA ▪ TH ▪ Ca ▪ Mg ▪ F ▪ Cl ▪ NO3 ▪ TC

Table 7.5: Water Quality Index (Rajpura Block)							
Parameter	Test Value	Standard Value	Deviation	Test Ratio	Water Quality Index	Indicator Value	Signal
pH	7.64	7.5	1.019	1.019	99.81	4	
TDS	726	500	1.452	1.452	95.48	4	
TA	277	200	1.385	1.385	96.15	4	
TH	363	300	1.210	1.210	97.90	4	
Ca	96.8	75	1.291	1.291	97.09	4	
Mg	29.4	30	0.980	1	100.00	5	
F	0.147	1	0.147	1	100.00	5	
Cl	110.9	250	0.444	1	100.00	5	
NO3-	7.407	45	0.165	1	100.00	5	
TC	198	10	19.800	11	0.00	1	
Overall Average					88.64	3	

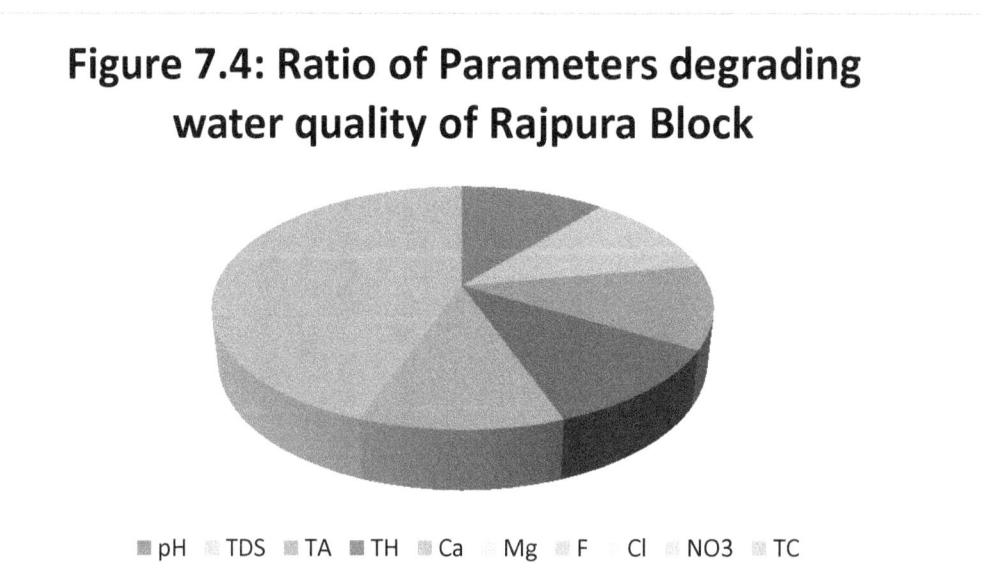

Figure 7.4: Ratio of Parameters degrading water quality of Rajpura Block

pH   TDS   TA   TH   Ca   Mg   F   Cl   NO3   TC

Table 7.6: Water Quality Index (Patiala Block)							
Parameter	Test Value	Standard Value	Deviation	Test Ratio	Water Quality Index	Indicator Value	Signal
pH	7.8	7.5	1.040	1.040	99.60	4	
TDS	912	500	1.824	1.824	91.76	3	
TA	349	200	1.745	1.745	92.55	3	
TH	456	300	1.520	1.520	94.80	4	
Ca	121.5	75	1.620	1.620	93.80	4	
Mg	36.9	30	1.230	1.230	97.70	4	
F	0.369	1	0.369	1	100.00	5	
Cl	139.6	250	0.558	1	100.00	5	
NO3-	3.057	45	0.068	1	100.00	5	
TC	82	10	8.200	8.200	28.00	1	
Overall Average					89.82	3	

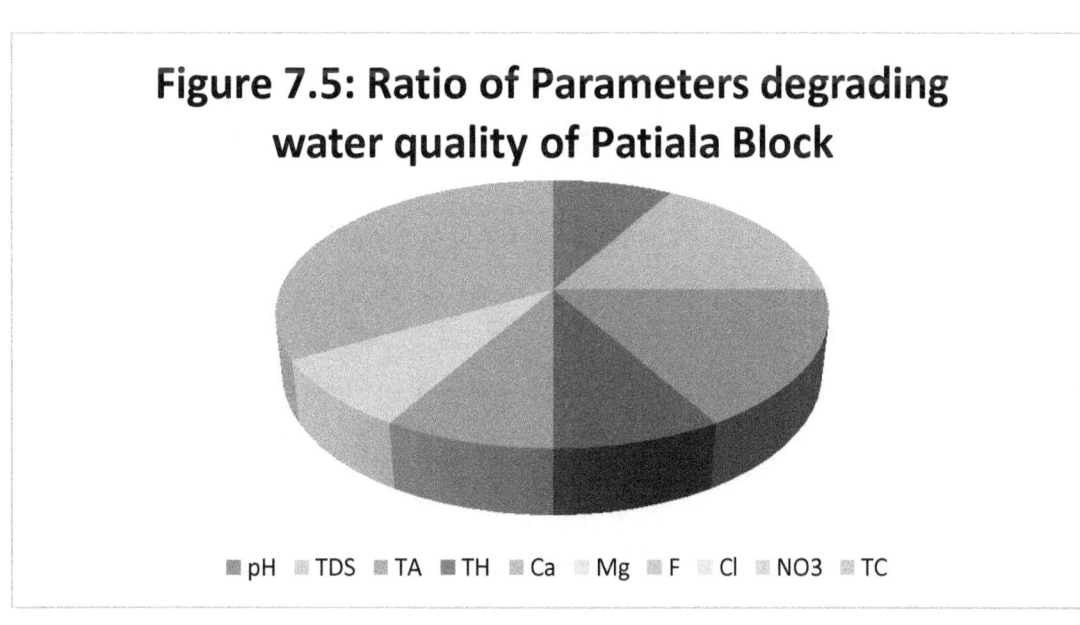

Figure 7.5: Ratio of Parameters degrading water quality of Patiala Block

■ pH ■ TDS ■ TA ■ TH ■ Ca ■ Mg ■ F ■ Cl ■ NO3 ■ TC

Table 7.7: Water Quality Index (Patran Block)							
Parameter	Test Value	Standard Value	Deviation	Test Ratio	Water Quality Index	Indicator Value	Signal
pH	8.05	7.5	1.073	1.073	99.27	4	
TDS	962	500	1.924	1.924	90.76	3	
TA	387	200	1.935	1.935	90.65	3	
TH	481	300	1.603	1.603	93.97	4	
Ca	128.3	75	1.711	1.711	92.89	3	
Mg	38.96	30	1.299	1.299	97.01	4	
F	0.39	1	0.390	1	100.00	5	
Cl	63.77	250	0.255	1	100.00	5	
NO3-	4.859	45	0.108	1	100.00	5	
TC	130	10	13.000	11	0.00	1	
Overall average					86.46	3	

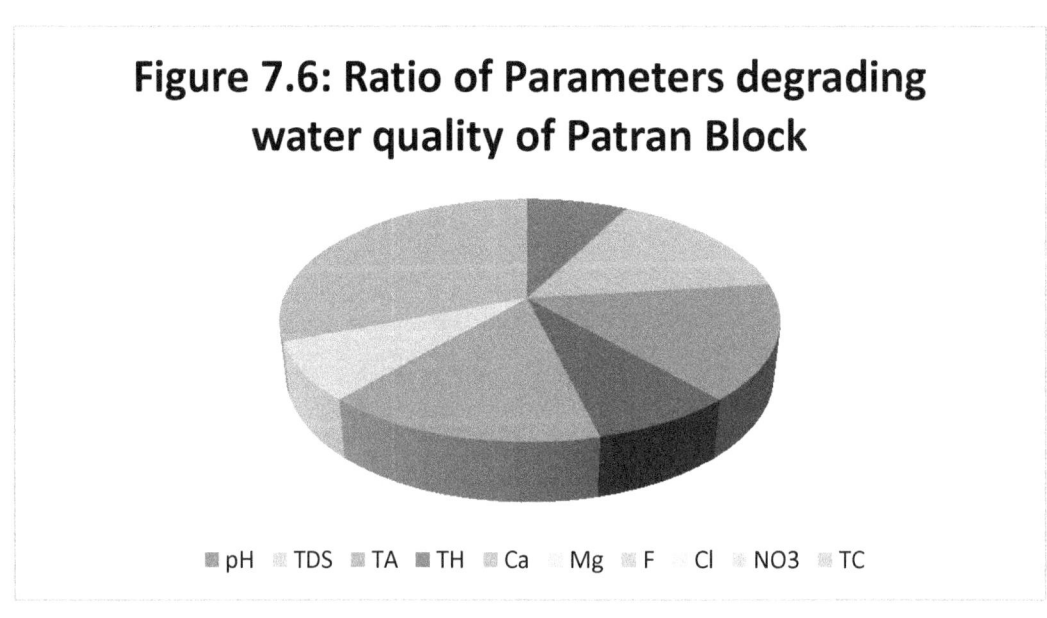

Figure 7.6: Ratio of Parameters degrading water quality of Patran Block

■ pH ▨ TDS ▨ TA ■ TH ▨ Ca   Mg ▨ F   Cl ▨ NO3 ▨ TC

Table 7.8: Water Quality Index (Samana Block)							
Parameter	Test Value	Standard Value	Deviation	Test Ratio	Water Quality Index	Indicator Value	Signal
pH	7.6	7.5	1.013	1.013	99.87	4	
TDS	805	500	1.610	1.610	93.90	4	
TA	305	200	1.525	1.525	94.75	4	
TH	403	300	1.343	1.343	96.57	4	
Ca	107.4	75	1.432	1.432	95.68	4	
Mg	32.61	30	1.087	1.087	99.13	4	
F	0.326	1	0.326	1	100.00	5	
Cl	56.7	250	0.227	1	100.00	5	
NO3-	2.023	45	0.045	1	100.00	5	
TC	54	10	5.400	5.400	56.00	2	
Overall Average					93.59	4	

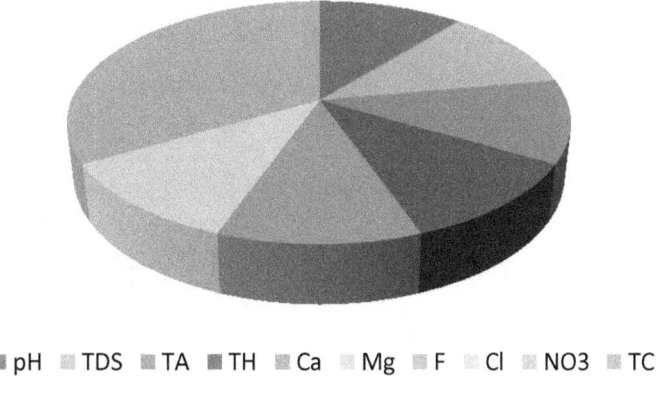

Figure 7.7: Ratio of Parameters degrading water quality of Samana Block

■ pH ■ TDS ■ TA ■ TH ■ Ca ■ Mg ■ F ■ Cl ■ NO3 ■ TC

Table 7.9: Water Quality Index (Sanour Block)							
Parameter	Test Value	Standard Value	Deviation	Test Ratio	Water Quality Index	Indicator Value	Signal
pH	8.15	7.5	1.087	1.087	99.13	4	
TDS	972	500	1.944	1.944	90.56	3	
TA	396	200	1.980	1.980	90.20	3	
TH	389	300	1.297	1.297	97.03	4	
Ca	69.5	75	0.927	1	100.00	5	
Mg	52.32	30	1.744	1.744	92.56	3	
F	0.262	1	0.262	1	100.00	5	
Cl	85.32	250	0.341	1	100.00	5	
NO3-	3.263	45	0.073	1	100.00	5	
TC	100	10	10.000	10.000	10.00	1	
Overall Average					87.95	3	

Figure 7.8: Ratio of Parameters degrading water quality of Sanour Block

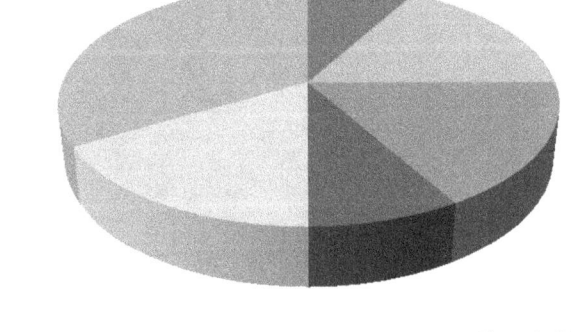

■ pH ▧ TDS ▨ TA ■ TH ▦ Ca Mg ▦ F Cl ▨ NO3 ▨ TC

Table 7.10: Water Quality Index (Patiala District as a Whole)							
Parameter	Test Value	Standard Value	Deviation	Test Ratio	Water Quality Index	Indicator Value	Signal
pH	7.93	7.5	1.057	1.057	99.43	4	
TDS	867	500	1.734	1.734	92.66	3	
TA	322	200	1.610	1.610	93.90	4	
TH	371	300	1.237	1.237	97.63	4	
Ca	79.3	75	1.057	1.057	99.43	4	
Mg	41.94	30	1.398	1.398	96.02	4	
F	0.338	1	0.338	1	100.00	5	
Cl	89.04	250	0.356	1	100.00	5	
NO3-	3.376	45	0.075	1	100.00	5	
TC	98	10	9.800	9.800	12.00	1	
Overall Average					89.11	3	

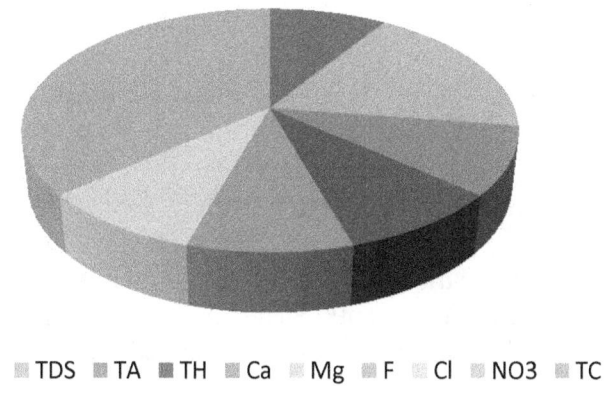

Figure 7.9: Ratio of Parameters degrading water quality of Patiala District

pH TDS TA TH Ca Mg F Cl NO3 TC

7.3   OUTPUT SIGNIFICANCE

The present research work may have the following significance:

➢ The assessment of drinking water quality and contamination can be useful for decision makers and daily human life, well being.
➢ Output of research can be informative for urban and rural water planning.
➢ "Save Water" and "Safe Water", slogans can be considered and implemented seriously.
➢ The need of the time, Sectoral Water Management and Integrated Water Management can be introduced and implemented.
➢ This research can be the basis of future research of drinking water quality and contamination using different spatial and temporal datasets.

7.4   LIMITATIONS OF THE STUDY

In this study, the following limitations were faced:

➢ For effective maintenance of water quality through appropriate control measures, continuous monitoring of large number of quality parameters is essential. However it is very difficult and laborious task for regular monitoring of all the parameters even if adequate manpower and laboratory facilities are available (Bhandari and Nayal, 2008). The limited number of water quality parameters studied in this research may bring out the limited reality of water.
➢ Convenient non-probability sampling method was employed to collect water samples which may not indicate the water quality of the total block or the district.
➢ Time constraint, by which any thesis is bound, may not be able to map the exact picture.

7.5   RECOMMENDATIONS:

The contamination of the drinking water is a serious problem, which needs to be addressed urgently. The available information regarding water purity can be utilized for policy formulation and managements of water use and water resources.

The ultimate goal of monitoring is to provide information and not data. In the past, many monitoring programmes have been characterized by the "data rich, information poor syndrome" (Ward et al., 1986). There should be more attention on the analysis and further use of collected data so that the end product of monitoring is information. If the data collected is not utilized in framing strategies and management, then the motive to collect data is lost. Recent developments in computing hardware and software have made it possible for a broader public to use data more effectively and obtain almost instantaneously results of simple data analysis. These technological and scientific improvements in recent years have not been institutionalized in many monitoring programs. Water quality data, today, are in many ways just lying about waiting to be combined, analyzed and interpreted in more meaningful and relevant ways for the public and managers.

Water monitoring is basically an information system. The science that serves as a basis for monitoring water systems is evolving rapidly and is necessarily broad and complex. It needs to cover the nature of environmental decision making, aquatic ecology, the statistics of analyzing data, the chemistry of water, the toxicity of chemicals to biological organisms, hydrology, data management hardware and software and many other areas of sciences.

There was a time when water was available in excess and pure and the users were less. There was no requirement of 'WATER MANAGEMENT' at that time. Then the time came when uses of water increased very much in large number of applications. It was the time to apply 'BASIC WATER MANAGEMENT'. With the many users and enormous use, there was the requirement of 'SECTORAL WATER MANAGEMENT'. Because of bypassing all that, we are now in a phase of increasing demands of water and leading shortage of quantity and good quality of water as well as serious damage to the environment. Now is the time to implement, 'INTEGRATED WATER MANAGEMENT' which includes 'differentiating between water use and water uses', 'sectoral water management' and 'policy management' before we enter into the time to implement 'COMPREHENSIVE WATER MANAGEMENT' which includes rationing and masking.

Strong pressure exerted by human activities on freshwater resources (Saeijs and van Berkel, 1995) has led to increased demands on policy and decision makers to develop well fundamental strategies and solutions. In water management and policy there is an increasing need for assessment methodologies for diagnosis and prognosis purposes in which an integrated water system approach is considered (van Rooy, 1995; Witmer, 1995). Now, the deterioration of water quality and environment has crossed the stage 'SCIENCE TO FIND' but there is a strong need to implement now 'SCIENCE OF ACTION'.

Based on the literature review, findings and analysis presented in this thesis, I offer the following recommendations to policy makers and future researchers:

7.5.1 For Policy Makers:

- Create Awareness among people who frame policies.
- Policies should originate from actual field surveys and authentic monitoring works.
- Proper urbanization of villages and villagers.
- In India and Punjab, many organizations public as well as private, are involved in water quantity and water quality issues. Result of that is miscommunication and unfavorable data quality. Government should create an independent and autonomous organization "WATER AUTHORITY" which should be responsible for all urgent water issues in one centralized body and recognize all other small organizations under law.
- There should be an effective and proper water management system for surface and groundwater.
- Proper and timely overhauling of the water distribution system.
- Organizations involved with water analysis should use new and advanced technology and tools for effective and accurate water analysis.
- While policy formulations and setting up Models regarding water issues, organizations in water issues should be involved actively.
- Use of datasets to cover longer and real time monitoring to ensure higher time frequency and more sampling points, should be encouraged.
- Wastewater drainage should be strengthened to get rid of pathogens and water related diseases.
- To frame policies and finding technology which mitigate the negative environmental impacts while at the same time producing sufficient and satisfying demands.
- Change the mindsets or habits of mind in favour of future adaptations and difficulties & problems.
- Water conservation, proper water use and recharge of water, are the need of hour. Aware everyone to have this motto as primary work of their life.

7.5.2 For Future Researchers:

The results of this study are not the end of this research. In future, there is need to update some of the data and improve the obtained results. For example:

- To include and study more water quality parameters related to surface and groundwater quality.
- Study to improve the detailed physical schematizations of the surface and groundwater networks and pollution outfalls.

- To extend the study of quality of water with reference to time and seasons.
- To collect and use more detailed data that represents time intervals and periods for greater accuracy and results.
- Refined and detailed spatial network and relation to rivers, hills and rocks.
- Elevation of cities and villages and comparison with air pressure effects, salinity and oxygen saturation.
- Study of water quality with reference to rainfall and snowmelt stream flows.
- Use of existent models to study the quality of water.
- Comparison of water quality models available and to create new models on water issues.
- Area wise water quality comparison to find better means to protect water quality.
- Role of Livestock to deteriorate water quality.
- To relate spatially human health with Water quality.
- Mathematical modeling to establish the relationship between water quality, livestock and human health.
- Comparison and effectiveness of various laws available regarding water issues.
- To find and identify the policies and technology which mitigate the negative environmental impacts while at the same time producing sufficient and satisfying demands.

Chapter-8

SUMMARY

Water is an absolute necessity of life as well as universal solvent. It contains dissolved materials and suspended particles even in its natural state. The quality of drinking water is rather a complex issue and involves various disciplines. Safe drinking water is a basic need of every human being despite of any socio-economic status. Drinking water criteria defines a quality of water that can be safely consumed by humans throughout their lifetime. Health hazards caused by unsafe water supplies are recognized as major problems in developing countries. Groundwater resources play a major role in ensuring livelihood security across the world, especially in economies that depend on agriculture. In India, majority of population is using groundwater for drinking and the contamination of this source is one of the major concerns due to unplanned urbanization, industrialization, over exploitation of the natural resources and discharge of hazardous wastes into water bodies without proper treatment. The magnitude of the problem of water availability and water quality is enormous, so the research problem regarding the quality of drinking water was considered to study the most confronting and current drinking water quality apprehensions and issues. The research problem "Physicochemical and microbiological quality assessment of rural drinking water supplies in district Patiala (Punjab)" was conducted in which two hundred drinking water samples from the villages of eight block of district Patiala were collected and assessed for various water quality parameters viz. temperature, pH, electrical conductivity, total dissolved solids, total suspended solids, total alkalinity, total hardness, calcium, magnesium, chloride, fluoride, nitrate and total coliforms.

Pollution of water originates from contamination of drinking water with domestic, industrial and agricultural sewage and wastes, causing the deterioration of both surface and groundwater. People are of the opinion that groundwater is safe and free from any kind of contamination. So, the people, especially the rural people do not feel the need of any kind of treatment for the groundwater, but poorly managed water supplies and water consumption have the potential to make a large number of people sick. It is not only the untreated water, but the treated stored water or the distribution system if managed improperly, may also prove fatal. Microbes can enter the water through soil by leaching, from the dumpsites or the sewerage system. Not only the human and livestock's faecal matter, but unplanned agriculture and unmanaged industrial waste can help in establishing unplanned and enormous growth of microbial colonies. Microbial growth can also occur due to unsatisfactory construction material coming into contact with water such as washers, pipe-lining compounds, and plastic used in pipes and taps. Corrosion, a common problem in water, can also be caused by chloride and microbes, besides iron, sulphates, nitrates and methane in association with microbes.

In the absence of governmental water supply or insufficient quantity, the people use hand-pumps or shallow bored tube-wells. Sewerage system and the blind pits made for drainage of waste water of households, affect the groundwater which is to be used from these sources of water. Over-abstraction of groundwater for agricultural activities and urbanization, have left not much underground water. Overexploitation is not only emptying the aquifers but also contaminating the water. The shortage and contamination of water is slowly affecting the lives of people as well as the environment around them. The water left underground, has impurities from the soil and rocks underneath, and is not very much suitable for drinking purposes. The underground water of shallow level has problems preferably of microbes and ions, and water of deep level is associated with heavy metals. Due to less availability and non-acceptance of surface water, people of Punjab and also Patiala, have to depend on groundwater resources to a greater extent. Even, in the many areas, groundwater is the main and only source of water. In rural areas, wells, hand-pumps, submersible pumps and tube-wells serve the main and fresh water source of drinking water.

For effective maintenance of water quality through appropriate control measures, continuous monitoring of large number of quality parameters is essential. However it is very difficult and laborious task for regular monitoring of all the parameters even if adequate manpower and laboratory facilities are available. The present study discuss the water quality parameters of general importance like pH, electrical; conductivity, total dissolved solids, total alkalinity, total suspended solids, total hardness, calcium, magnesium, fluoride, chloride, nitrate and total coliforms. Convenient non-probability sampling method and cluster sampling methods were used to collect samples. Prescribed and standard methods were used to analyze water samples collected. The results were studied and discussed block-wise as well as group and cluster-wise.

The total area of Patiala district can be divided into two clusters. Cluster one may comprise, Bhunerheri, Ghanaur, Patiala, Sanour and Patran from where Ghaggar River and Patiale-wali Nadi passes. These rivers carry waste water because of waste dumped by industries and municipalities. Cluster two may comprise the other area, Rajpura, Nabha and Samana. Table number 5.22 to 5.25 include the mean value, standard deviations and correlation coefficients of the cluster-1 and cluster-2 and Figure No. 5.28 & 5.29 compare the water quality parameters of these two clusters. Study regarding water quality parameters of Patiala district as a whole is also studied (Table No. 5.26, 5.27). Table No. 7.2 to 7.10 and Figure No. 7.1 to 7.9 indicate the pictures of the quality of water of eight blocks of Patiala district.

Two quality parameters, Total Dissolved Solids (Table No.5.19 and Figure No.5.30 and Total Hardness (Table No.5.18 and Figure No.5.2) are of general importance. On the basis of physicochemical monitoring of the samples studied, it may be concluded that the water of all the samples studied is of hard type and none of the samples showed soft or moderately hard type water quality. Water with Total Hardness below 75 can be considered soft, 75 to 150 moderately hard, 151 to 300 hard and water with Total Hardness above 300 is termed as very hard (Matalas and Reiher, 1967 & Sawyer and McCarthy, 1967). Table No. 5.18 and Figure No. 5.2 indicates that 80% is hard water and 20% of water is very hard in the Bhunerheri block. Ghanaur block has 67% hard water and 33% very hard water. Water of Nabha block can be classified as 50% hard and 50% very hard. Rajpura block has 44% hard and 56% very hard water. Patiala block has 25% hard and 75% very hard water. Patran block contains 25% hard and 75% very hard water. Water of Samana block can be classified as 20% hard and 80% very hard. Sanour block contains 10% hard water and 90% very hard water. Consumption of very soft and very hard water is not good for human beings. Very soft water does not contain much solids dissolved, needed to digest the food while very hard water contains much solids dissolved in it and can create problems. So the water samples come under Slightly Hard and Moderately Hard category, can be consumed safely, if otherwise found suitable. As per Total Dissolved Solids are concerned, safe and desirable limit as prescribed by BIS and WHO is 500, but water with Total Dissolved Solids up to 1500 (WHO) and 2000 (BIS) can be consumed if no other alternative is available. On the basis of physicochemical monitoring of the samples studied (Table No.5.19 and Figure No.5.3), it may be concluded that none of the samples showed Total Dissolved Solids below 300 means no water can be termed as good. Water with Total Dissolved Solids 300 to 500 can be considered fair, 500 to 900 average, 900 to 1200 poor, 1200 to 2000 very poor and water with Total Dissolved Solids above 2000 is termed as unacceptable (Matalas and Reiher, 1967). On the basis of Total Dissolved Solids, it was found that none water samples in Bhunerheri block, 17% water samples in Ghanaur block, 44% water samples in Nabha block, 15% water samples in Patran block, 25% water samples in Patiala block, 33% water samples in Rajpura block, 10% water samples in Samana block and 5% water samples in Sanour block may be termed as fair. Average plus poor category include 80% water samples in Bhunerheri block, 70% water samples in Ghanaur block, 44% water samples in Nabha block, 60% water samples in Patran block, 60% water samples in Patiala block, 57% water samples in Rajpura block, 85% water samples in Samana block and 75% water samples in Sanour block. All the block include some water samples in very poor category and one water samples was found in Ghanaur block in unacceptable category.

The overall quality of drinking water as rated by the water quality index and indicator value for Bhunerheri block is 3 (Satisfy Function Needs Minimal),

Ghanaur block as 3 (Satisfy Function Needs Minimal), Nabha block as 4 (Satisfy Function Needs Optimal), Rajpura block as 3 (Satisfy Function Needs Minimal), Patiala block as 3 (Satisfy Function Needs Minimal), Patran block as 3 (Satisfy Function Needs Minimal), Samana block as 4 (Satisfy Function Needs Optimal), and Sanour block as 3 (Satisfy Function Needs Minimal). The drinking water quality of Patiala district as a whole can be rated as 3 (Satisfy Function Needs Minimal) on the basis of water quality index and indicator value.

It can be concluded that the water from maximum villages of all the blocks of district Patiala is not suitable for drinking; even sufficient quantities are not available. Disinfection and treatment of water is suggested before drinking. If proper care and management to ascertain quantity and quality is not applied, not the quantity but the quality of the drinking water will be deteriorated to the much extent. Very shortly a situation may come when not only the insufficient quantity but that too highly impure, will be left with us to offer to our offspring.

Adriano D C, **1986**, Trace elements in the terrestrial environment, *Springer – Verlag*, New York, 533p.

Agnihotri N P, **2000**, Pesticide Consumption in Agriculture in India- An update, *Pesticide Research Journal*, **12(1)**, 150-55.

Agrawal V, Vaish A K and Vaish P, **1997**, Groundwater quality: Focus on fluoride and fluorosis in Rajasthan, *Current Science*, **73**, 743-46.

Ahmad A and Alam M, **2003**, Physicochemical and toxicological studies of industrial effluents in and around Delhi and ground water quality of some areas in Delhi city, *Chemical Environment Research*, **12(1&2)**, 5-13.

Amathussalam A, Abubacker M N, Jayabal N, **2002**, Impact of sugar mill effluent on ground water - a case study, *Journal of Industrial Pollution Contamination*, **18(2)**, 119-24.

Anaissie E J, Penzak S R, Dignani M C, **2002**, The hospital water supply as a source of nosocomial infection: a plea for action, *Archives of Internal Medicine*, **162(13)**, 1483-92.

Anand C, Akolkar P, Chakrabarti R, **2006**, Bacteriological water quality status of river Yamuna in Delhi. *Journal of Environmental Biology*, **27(1)**, 97-101.

Ansaruzzaman M, Albert M J, Nahar S, Byun R, Katouli M, Kuhn I, Mollby R, **2000**, Clonal group of enteropathogenic E. Coli isolated in case control studies of diarrhea in Bangladesh, *Journal of Medical Microbiology*, **49**, 177-85.

APHA (American Public Health Association), **1989**, Standard Methods for examination of water and waste water, *American Public Health Association*, Vol. **15**, Washington, DC.

APHA, **1995**, Standard Methods for examination of water and waste water, **18**[th] Ed., *American Public Health Association*, Washington, DC.

Ashour A J, Hung L, **2000**, Transport of bacteria on sloping soil surfaces by runoff, *Environmental Toxicology*, **15(2)**, 149-53.

ASTM (American Society for Testing and Materials), **1972**, *Annual Book of ASTM Standards*, ASTM, Philadelphia, USA.

Atef A, Al-Kharabsheh, **1999**, Influence of urbanization on water quality at Wadi Kufranja basin (Jordan), *Journal of Arid Environments*, **43(1)**, 79-89.

Au J, Bagchi P, Chen B, Martinez R, Dudley S A, Sorger G J, **2000**, Methodology for public monitoring of total coliforms, E. Coli and toxicity in waterways by Canadian high school students, *Journal of Environmental Management*, **58(3)**, 213-30.

AWWA, **1990**, Water quality and treatment: a handbook of community water supplies, 4th Ed., *McGraw Hill*, Inc. USA.

Ayoob S and Gupta A K, **2006**, Fluoride in Drinking Water: A Review on the Status and Stress Effects, *Critical Reviews in Environmental Science and Technology*, **36(6)**, 433-87.

Bajwa M S, Singh B, and Singh P, **1993**, Nitrate Pollution of Groundwater under Different Systems of Land Management in Punjab, (Narain P, Ed., *Proceeding of First Agriculture Science Congress*, National Academy of Agriculture Sciences, New Delhi), 223-30.

Baloch M K, Jan I U, Ashour S T, **2000**, Effects of septic tank effluents on quality of groundwater, *Pakistan Journal of Food Science*, **10(3-4)**, 25-31.

Barik R N, Pradhan B, Patel R K, **2005**, Trace elements in ground water of Paradip area. *Journal of Industrial Pollution Contamination*, **21(2)**, 355-62.

Baroni L, Cenci L, Tettamanti M, Berati M, **2007**, Evaluating the environmental impact of various dietary patterns combined with different food production systems, *European Journal of Clinical Nutrition*, **61**, 279-86.

Battu R S, **2005**, Status of pesticide residues in Talwandi Sabo, Bathinda, Report on survey conducted by *Department of Entomology and Director of Research*, Punjab Agricultural university, Ludhiana.

Battu R S, Chawla R P and Kalra R L, **1978**, Insecticide Residues in Market Samples of Vegetables Oils and Oilseeds Cakes from selected area of Punjab, *Indian Journal of Ecology*, **7(1)**, 1-8.

Betancourt Q Q W, De-Ledesma L B, **2000**, Descriptive study on the presence of protozoan cysts and bacterial indicators in a drinking water treatment plant in Maracaibo, Venezuela, *International Journal of Environmental Health Research*, **10(1)**, 51-61.

Bhandari N S, Nayal K, **2008**, Correlation Study on Physicochemical Parameters and Quality Assessment of Kosi River Water, Uttarakhand, *E-Journal of Chemistry*, **5(2)**, 342-46.

Bhatt R, **2010,** Water resource management for sustainable crop production in India, *Agrobios*, **VIII,** 15-17.

Booth D B, and Jackson C R, **1997**, Urbanization of aquatic systems: Degradation thresholds, storm-water detection and the limits of mitigation, *Journal of the American Water Resources Association*, **33(5)**, 1077-90.

Boulay N, Edwards M, **2001**, Role of temperature, chlorine and organic matter in copper corrosion by-product release in soft water, *Water Research*, **35(3)**, 683-90.

Brar J S and Chibba I M, **1997**, Potassium status and quality of underground water of Punjab, *Indian Journal of Ecology*, **24(2)**, 165.

Brar M S, Khurana M P S and Kansal B D, **2002**, Effect of irrigation by untreated sewage effluents on the micro and potentially toxic elements in soils and plants, In *Proc. 17th World Congress of Soil Science held at Bangkok*, Thailand from August 14-21, 2002, **IV(24)**, p198.

Brar S P S, Kumar D and Bishnoi S R, **1984**, Hydrochemistry of underground waters of Bhawanigarh block (Sangrur district), *Indian Journal of Environmental Health*, **23(3)**, 202-11.

Briscoe J and Malik, R P S, **2006**, India's Water Economy: Bracing for a Turbulent Future, The World Bank, Oxford University Press, New Delhi.

Bunn S E, Arthington A H, **2002,** Basic principles and ecological consequences of altered flow regimes for aquatic biodiversity, *Environmental Management*, **30(4)**, 492–507.

Burch J and Thomas K, **1998**, Water disinfection for developing countries and potential for solar thermal pasteurization, *Solar Energy*, **64(1-3)**, 87-97.

Chakraborti D, Biswas B K, Basu G K, Mandal B K, Chowdhury U K, Mukherjee S D, Gupta J P, Chowdhury S R and Rathore K C, **1999**, Arsenic groundwater contamination and sufferings of people in Rajnandgaon district, Madhya Pradesh, India, *Current Science*, **77**, 502 - 04.

Chang A C, Peng C, Page A L, Asano T, **2002**, Developing human health related chemical guidelines for reclaimed waste and sewage sludge applications in agriculture, World Health Organization, Geneva, p.105.

Chapman D, **1996**, *Water Quality Assessments*, World Health Organization, p113.

Charlet L and Polya D A, **2006**, Arsenic in shallow, reducing ground waters in southern Asia: an environmental health disaster, *Elements*, **2**, 91-96.

Charriere G, Mossel D A A, Beaudeau P, LeClere H, **1994**, Assessment of the marker value of various components of the coli-aerogens group of enterobacteriaceae and of a selection of enterococcus species for the official monitoring of drinking water supplies, *Journal of Applied Bacteriology*, **76(4)**, 336-44.

Chattopadhyay P K, **1998**, Insecticide and Pesticide Pollution of Food Stuffs and their Toxic Effect on Man, Project sponsored by *Punjab State Council for Science and Technology*, Chandigarh.

Chilton P J, Stuart M E, Escolero O, Marks R J, Gonzalez A, Milne C J, **1998**, Groundwater recharges and pollutant transport beneath wastewater irrigation: the case of Leon, Mexico, p153-68, in *Groundwater Pollution, Aquifer Recharge, and Vulnerability*, Geological Society, London.

Choubisa, S.L., **1998**, Fluorosis in some tribal villages of Udaipur district (Rajasthan), *Journal of Environmental Biology*, **19**, 341-52.

Clark S C, Lawler D F and Cushing R S, 2004, Contact filtration: particles and ripening, *Journal of the American Water Works Association*, **84**, 61–71.

Clement M, Seux R, Rabarot S, **2000**, A practical model for estimating total lead intake from drinking water, *Water Research*, **34(5)**, 1533-42.

Codd G A, Ward C J, Bell S G, **1997**, Cyanobacterial toxins: occurrence, modes of action, health effects and exposure routes, *Archives of Toxicology*, Suppl., **19**, 399-410.

Crabill C, Donald R, Snelling J, Foust R, Southam G, **1999**, The impact of sediment faecal coliform reservoirs on seasonal water quality in Oak Creek, Arizona, *Water Research*, **33(9)**, 2163-71.

Dangendorf F, Herbst S, Reintjes R, Kistemann T, **2002**, Spatial patterns of diarrheal illness with regard to water supply structures- A GIS analysis, *International Journal of Hygiene and Environmental Health*, **205(3)**, 183-91.

Das R, Samal N R, Roy P K, Mitra D, **2006**, Role of electrical conductivity as an indicator of pollution in shallow lakes, *Asian Journal of Water Environment Pollution*, **3(1)**, 143-46.

Datta P S, Tyagi S K, Mookerjee P, Bhattacharya S K, Gupta N and Bhatnagar P D, **1999**, Groundwater NO_3 and F contamination processes in Puskar valley, Rajasthan as reflected from ^{18}O isotopic signature and ^{3}H recharge studies, *Environment Monitoring and Assessment*, **56**, 209-19.

Daw R K, **2004**, Experiences with domestic defluoridation in India, *Proceedings of the 30th WEDC International Conference on People-Centered Approaches to Water and Environmental Sanitation*, Vientiane, Lao PDR, 467–73.

Dayal G, **1992**, Groundwater qualities of rural and urban settle mentsat Agra, *Journal of Nature Conservation*, **4(1)**, 89-93.

Dhillion S K and Dhillion K S, **1997**, *Land Conservation and Reclamation*, **5**, 313-23.

Dhillon K S, Dhillon S K and Singh K P, **2004**, Selenium Toxicity in Punjab, Special Report, *Punjab State Council of Science and Technology*, p24.

Dixit RC, Verma S R, Nitnaware V, Thacker N P, **2003**, Heavy metals contamination in surface and groundwater supply of an urban city, *Indian Journal of Environment Health*, **45(2)**, 107-12.

Dodds W K, **2002**, Freshwater Ecology, Concepts and Environmental Applications, Academic Press, USA.

Dojlodo J R, Best G A, **1993**, Chemistry of Water and Water Pollution, Ellis Harwood Ltd., Great Britain.

Dutta R K, Saikia G, Das B, Bezbaruah, Das H B, Dube S N, **2006**, Fluoride contamination in groundwater of Central Assam, India. *Asian Journal of Water Environment Pollution*, **3(2)**, 93-100.

Easton Z M, Gerard-marchant P, Walter M T, Petrovic A M, Steenhuis T S, **2007**, Identifying dissolved phosphorus source areas and predicting transport from an urban watershed using distributed hydrologic modeling, *Water Resources Research*, **43(11)**, 1-16.

Edberg S C, Le Clere H, Robertson J, **1997**, Natural protection of spring and well drinking water against surface microbial contamination-II, indicators and monitoring parameters parasites, *Critical Reviews in Microbiology*, **23(2)**, 179-206.

Ellis K, **1991**, Water disinfection: a review with some consideration of the requirements of the third world, *Critical Reviews of Environmental Control*, **20(5&6)**, 341-407.

Ermosele C O, Ermosele I C, Muktar S A and Birdling S A, **1995**, Metals in fish from the upper Benue River and Lakes Geryo and Njuwa in Northern Nigeria, *Bulletin of Environmental Contamination and Toxicology*, **54**, 8-14.

ESCAP, **1994**, Guidelines on monitoring methodologies for water, air and toxic chemicals, hazardous waste, *Economic and Social Commission for Asia and Pacific* (ESCAP), United Nations, New York.

ESCAP, **1995**, Integrated water resources management in Asia and the Pacific, Water resources series no. 75, *Economic and Social Commission for Asia and Pacific* (ESCAP), United Nations, New York.

ESCAP, **1997**, Guidelines on Water and Sustainable Development: Principles and Policy Options, Water resources series no. 77, *Economic and Social Commission for Asia and Pacific* (ESCAP), United Nations, New York.

Eynard F, Mez K, Walther J L, **2000**, Risk of cyanobacterial toxins in Riga waters (Latvia), *Water Research*, **34(11)**, 2979-88.

Falcao J P, Dias A M G, Correa E F, Falcao D P, **2002**, Microbiological quality of ice used to refrigerate foods, *Microbiology*, **19(4)**, 269-76.

Farah N, Zia M A, Rehman K and Sheikh M, **2002**, Quality characteristics and treatment of drinking water of Faisalabad city, *International Journal of Agricultural Biology*, **3**, 347–49.

Farooq M A, Malik M A, Hussain A, Abbasi H N, **2010**, Multivariate Statistical Approach for the Assessment of Salinity in the Periphery of Karachi, Pakistan, *World Applied Sciences Journal,* **11(4),** 379-387.

Fotedar A, Verma R K and Fotedar B K, **2008**, Physicochemical Studies of the Water-bodies in and Around Shivkhori Area, Jammu Himalaya, in Relation to

Geology of the Area, *Nature Environmental and Pollution Technology*, **7(3)**, 489-99.

Fraiture C de and Wichelns D, **2010**, Satisfying future water demands for agriculture, *Agricultural Water Management*, **97**, 502–11.

Framen M A, **1981**, Standard methods of examination of the water and wastewaters, **15**[th] edition, *American Public Health Association*, Washington DC.

Frick E A, Buell G R, and Hopkins E H, **1996**, Nutrient sources and analysis of nutrient water-quality data, *Water Resources Investment*, (Rep.), 96-4101, USGS, Reston, VA.

Friedl G, Teodoru C and Wehrli B, **2004**, Is the Iron Gate I reservoir on the Danube River a sink for dissolved silica?, *Biogeochemistry*, **68**, 21-32.

Funari E, Ottaviani M, **1997**, Hygiene and health effects of drinking water, *National Technical Information Service-Report*, USA, **97(9)**, p180.

Gardner-Outlaw T and Engleman R, **1997**, Sustaining Water, Easing Scarcity: A Second Update, *Population Action International*, Washington DC.

Garg V K, Dahiya S, Chaudhary A and Shikha D, **1998a**, Fluoride distribution in ground waters of Jind district, Haryana, India, *Ecology Environment and Conservation*, **4(1-2)**, 19-23.

Garg V K, Sharma I S and Bishnoi M S, **1998b**, Fluoride in underground waters of Uklana town, district Hisar, Haryana, *Pollution and Research*, **17(2)**, 149-52.

Geldreich E E, **1991**, Microbial water quality concerns for supply use, *Environment Toxicology and Water*, **6**, 209-23.

Geological Survey of India (GSI), **1963**, Annotated index of Indian mineral occurrences-Part (II), Edited by Chaterjee P K, p.147 - 285.

General guidelines for water audit and water conservation, **2005**, Central water commission, *Ministry of water resources*, Government of India.

Genthe B N, Strauss, **1997**, The effect of type of water supply quality in a developing community in South Africa, *Water Science Technology*, **35(11–12)**, 35–40.

Germani Y, Morillon M, Begaud E, Dubourdieu H, Costa R, and Thevenon I, **1994**, Two year study of endemic enteric pathogens associated with acute diarrhea in New Caledonia, *Journal of Clinical Microbiology*, **32(6)**, 1532-36.

Gleick P, **1999**, The Human Right to Water, *Water Policy*, **1(5)**, 487-503.

Gordon L J, Finlayson M, Falkenmark M, **2010,** Managing water in agriculture for food production and other ecosystem services, *Agricultural Water Management*, **97**, 512–19.

Government of India, **2007**, Report of the Expert Group of Groundwater management and Ownership, *Planning Commission*, New Delhi.

Goyal M R, Abrol O P, Vohra A K, **1981**, Pollution of upper aquifer in Punjab (India), *Studies in Environmental Science*, **17**, 105-10.

Grabow W, **1991**, Human virus in water, *Water Sewage Effluents*, **11**, 16-21.

Grant M A, **1997**, A new membrane filtration medium for simultaneous detection and enumeration of Escherichia coli and total coliforms, *Applied Environmental Microbiology*, **63**, 3326-530.

Gulfraz M, Afjal H, Malik M A, Asrar M, Hayat M A, **1997**, A study of water pollution caused by the effluents of various industries located in the vicinity of Sohan River, *Pakistan Journal of Science*, **46**, 1-2.

Gulis G, Czompolyova G, Cerhan J R, **2002**, An ecological study of nitrate in municipal drinking water and cancer incidence in Trnava district, Slovakia, *Environment Research*, **88(3)**, 182-87.

Gupta D, Deswal S, Kumar D and Das G, **2009**, Assessment of Water Quality in the Villages of Ambala District, Haryana (India), *Proceedings of International Conference on Energy and Environment*, March 19-21, 358-62.

Gupta M K, Singh V, Rajwanshi P, Agarwal M, Rai K, Srivastva S, Srivastva R and Dass S, **1999**, Ground water quality assessment of tehsil Kheragarh, Agra (India) with special reference to fluoride, *Environment Monitoring and Assessment*, **59**, 275-85.

Guru Prasad B, **2003**, Evaluation of water quality in Tadepalli mandal of Guntur district, AP, *Nature Environmental Pollution and Technology*, **2(3)**, 273-76.

Hacioglu N and Dulger B, **2009**, Monthly variation of some physicochemical and microbiological parameters in Biga Stream (Biga, Canakkale, Turkey), *African Journal of Biotechnology*, **8(9)**, 1929-37.

Hamzah A, Abdullah M P, Sarmani S, Johari M A, **1997**, Chemical and bacteriological monitoring of drinking water from an urbanized water catchment drainage basin, *Environmental Monitoring and Assessment*, **44**, 327-28.

Harrison W N, Bradberry S M, Vale J A, **2000**, Chemical contamination of private drinking water supplies in the West Midlands, UK, *Journal of Toxicology: Clinical Toxicology*, **38(2)**, 137-44.

Hegde S N and Puranik S C, **1992**, Trace elements in groundwater of Hubli city, Karnataka, India, *Current Science*, **63(1)**, 43-45.

Hem J D, **1970**, Study and Interpretation of the Chemical Characteristics of Natural Water, Second Edition-*Geological Survey Water Supply Paper* 1473, US Government Printing Office, Washington.

Hira G S, Jalota S K and Arora V K, **2004,** Efficient management of water resources for sustainable cropping in Punjab. *Research Bulletin: Department of Soils*, PAU, Ludhiana.

Hira G S, Singh R and Kukal S S, **2002**, Soil metric suction: a criterion for scheduling irrigation to rice, *Indian Journal of Agricultural Sciences*, **72,** 236-37.

Holgate G, **2000**, Water quality: DETR consultation on new regulations for drinking water, *Environment and Waste Management*, **3(3)**, 105-12.

Holt M S, Eisenbrand G, Hofer M, Kroes R, Shuker I, **2000**, Sources of chemical contaminants and routes into the fresh water environment, *Food and Chemical Toxicology*, **38**, S21-27.

Hrudey S E, Payment P, Huck P M , Gillham R W and Hrudey E J, **2003**, A fatal waterborne disease epidemic in Walkerton, Ontario: comparison with other waterborne outbreaks in the developed world, *Water Science and Technology*, **47(3)**, 7–14.

Huang G H, Xia J, **2001**, Barriers to sustainable water quality management, *Journal of Environmental Management*, **61(1)**, 1-23.

Huang W Y, Beach E D, Fernandezeornejo J, **1994**, An assessment of the potential risk of groundwater and surface water contamination by agricultural chemicals used in vegetable production, *Science of the Total Environment*, **153(1-2)**, 151-67.

Hundal H S, Kumar R, Singh K and Singh D, **2007**, Occurrence and Geochemistry of Arsenic in Groundwater of Punjab, *Soil Science and Plant Analysis*, **38 (17&18)**, 2257-77.

Hunter P R, Colford J M, LeChevallier M W, Binder S and Berger P S, **2001**, Waterborne Diseases, U.S. Environmental Protection Agency, *Emerging Infectious Diseases*, **7(3)**, 544-55.

Hussain I, Hanjra M A, **2004**, Irrigation and poverty alleviation, review of the empirical evidence, *Irrigation and Drainage*, **43**, 1-15.

Indra V and Sivaji S, **2006**, Metals and organic components of sewage and sludges, *Journal of Environmental Biology*, **27**, 723-25.

Islam S R, Gyananath G, **2002**, Contamination of chemical fertilizers in groundwater, *Journal of Ecotoxicology and Environment Monitoring*, **12(4)**, 285-90.

Iwami O, Watanabe T, Moon C S, Nakatsuka H, Ikeda M, **1994**, Motor neuron disease on the Kii Penninsula of Japan – excess manganese intake from food coupled with low magnesium in drinking water as a risk factor, *Science of the Total Environment*, **149(1-2)**, 121-35.

Iyer R R, **2003**, Water: Perspectives, issues, concerns, Sage Publications India Pvt. Ltd., New Delhi.

Jacinthe P A, Dick W A, Brown L C, **2000**, Bioremediation of nitrate contaminated shallow soils and waters via water table management techniques: evolution and release of nitrous oxide, *Soil Biology and Biochemistry*, **32(3)**, 371-82.

Jadeja B A, Odedra N K, Thaker M R, **2006**, Studies on ground water quality of industrial area of Dharampur, Porbandar city, Saurashtra, Gujrat, India. *Plant Archives*, **6(1)**, 341-44.

Jakher G R and Rawat M, **2003**, Correlation of nitrate and most probable number for a sewage fed pond, Gulab Sagar at Jodhpur city, *Oikoassay*, **16(1)**, 13-14.

Janmaat J, **2004**, Calculating the cost of irrigation induced soil salinization in the Tungabhadra project, *Agricultural Economics*, **31**, 81-96.

Jardine C G, Gibson N, Hrudey S E, **1997**, Detection of odour and health risk perception of drinking water, in *Off Flavours in the Aquatic Environment*, Elsevier Science Ltd., Pergamon, UK, 91-98.

Jena B, Sudarshana R, Chaudhury S B, **2003**, Studies on water quality parameters around Sagar Island, Sundarbans, *Nature, Environment Pollution Technology*, **2(3)**, 329-32.

Joia B S, Chawla R P and Kalra R L, **1978**, Residue of DDT and HCH in wheat flour in Punjab, *Indian Journal of Ecology*, **5(2)**, 120-27.

Jorgensen A, Nohr K, Sorensen H, Boisen F, **1998**, Decontamination of drinking water by direct heating in solar panels, *Journal of Applied Microbiology*, **85**, 441-47.

Kalra R L and Chawla R P, **1980**, Occurrence of DDT and BHC Residues in Human Milk in India, *Experientia*, **37**, 404-05.

Karbassi A R, Nouri J, Mehrdadi N, Ayaz G O, **2008**, Flocculation of heavy metals during mixing of freshwater with Caspian Sea water, *Environment Geology*, **53(8)**, 1811-16.

Karim M M, **2000**, Arsenic in groundwater and health problems in Bangladesh, *Water Research*, **14**, 1304–10.

Kaur A, Pallah B S, Sahota G P S, Sahota H S, **1992**, Seasonal variability of chemical parameters in drinking water from shallow aquifers, *Indian Journal of Environment Protection*, **12(6)**, 409-15.

Kaushik A, Jain S, Dawra J, Sharma P, **2003**, Heavy metal pollution in various canals originating from river Yamuna in Haryana, *Journal of Environmental Biology*, **24(3)**, 331-37.

Kaushik A, Kumar K, Sharma K, **2002**, Water quality index and suitability assessment of underground water of Hisar and Panipat in Haryana, *Journal of Environmental Biology*, **23**, 325-33.

Kent R H and Belitz K, **2004**, Concentrations of dissolved solids and nutrients in water sources, *Water Resources Investment* (Rep.), 03-4326, USGS, Reston, VA.

Khan S, Tariq R, Yuanlai C, Blackwell J, **2006,** Can irrigation be sustainable?, *Agriculture Water Management*, **80(1–3)**, 87–99.

Khedkar D D and Dixit A J, **2003**, Physicochemical analysis of domestic wastewater of Amravati (Maharashtra), *Journal of Aquatic Biology*, **18(1)**, 69-72.

Khurana M P S and Bansal R L, **2008**, Impact of sewage irrigation on speciation of nickel in soils and its accumulation in crops of industrial towns of Punjab, *Journal of Environmental Biology*, **29(5)**, 793-98.

Khurana M P S, Nayyar V K, Bansal R L and Singh M V, **2003**, Ground Water Pollution (Singh V P and Yadav R N eds.), Allied Publishers Pvt. Ltd, New Delhi, 487-95.

Kim G, Shon J, Won H, Hyun J, Oh W, **2002**, A study of methods to reduce groundwater contamination around the Kimpo landfill in Korea, *Environmental Technology*, **23(5)**, 561-70.

KnobelochL, Salna B, Hogan A, Postle J, Anderson H, **2000**, Blue babies and nitrate contaminated well water, clinical reference, *Environmental Health Perspectives*, **108(7)**, 675-78.

Knox R C, Canter L W, **1996**, Prioritization of groundwater contaminants and sources, *Water, Air, Soil Pollution*, **88(3-4)**, 205-26.

Kolpin D W, Furlong E T, Meyer M T, Thurman E M, Zaugg S D, Barber L B, and Buxton H T, **2002**, Pharmaceuticals, hormones, and other organic wastewater contaminants in U.S. streams, 1999-2000: a national reconnaissance, *Environmental Science and Technology*, **36**, 1202-11.

Koppe P, **1973**, Suggestions regarding control of water bodies – groundwater, *Bull. Inf. Fed. Eur. Prot. Eaux.*, **20**, 71-75.

Krishnamurthy R, **1996**, Water in Ancient India, *Indian Journal of History of Science*, **31(4)**, 327-37.

Krishnan R R, Dharmaraj K and Ranjitha Kumari B D, **2007**, A comparative study on the physicochemical and bacterial analysis of drinking, bore well and sewage water in the three different places of Sivakasi, *Journal of Environmental Biology*, **28**, 105-08.

Krishnaswami S and Singh S K, **2005**, Chemical weathering in the river basins of the Himalaya, India, *Current Science*, **89**, 841–49.

Kumar N, Sinha D K, **2010**, Drinking water quality management through correlation studies among various physicochemical parameters: A case study, *International Journal of Environmental Sciences*, **1(2)**, 253-59.

Kumar R, **2005**, An Epidemiological study of cancer cases reported from villages of Talwandi Sabo, Bathinda, *Project Report, PGIMER*, Chandigarh and PPCB, Patiala.

Kumar S, **2008**, Studies on ground water pollution from dumping of municipal solid wastes at Muzaffarpur, *Journal IHHE India*, 2008-09**(4)**, 14–18.

Kumari B S, Kavitha Kirubavathy A and Thirumalnesan R, **2006**, Suitability and water quality criteria of an open drainage municipal sewage water at Coimbatore, used for irrigation, *Journal of Environmental Biology*, **27**, 709-12.

Lampman W, **1995**, Susceptibility of groundwater to pesticide and nitrate contamination in predisposed areas of southwestern Ontario, *Water Quality Research Journal of Canada*, **30**, 443-68.

Landon M K, Delin G N, Komor S C, Regan C P, **2000**, Relation of pathways and transit times of recharge water to nitrate concentrations using stable isotopes, *Ground Water*, **38(3)**, 381-95.

Langston W J, Burt G R and Pope N D, **1999**, Bioavailability of Metals in Sediments of the Dogger Bank (Central North Sea): A Mesocosm Study, *Estuarine, Coastal and Shelf Science*, **48**, 519-40.

Latour P J M, Stutterheim E and Schafer A J, **1996**, From data to information: the WATER DIALOGUE, H_2O, **29**, 693-96.

Lawson N M, Mason R P, **2001**, Concentration of mercury, methyl mercury, cadmium, lead, arsenic and selenium in the rain and stream water of two contrasting watersheds in Western Maryland, *Water Research*, **35(17)**, 4039-52.

Lee S H, Kim S J, **2002**, Detection of infectious enteroviruses and adenoviruses in tap water in urban areas in Korea, *Water Research*, **36(1)**, 248-56.

Lehloesa L J and Muyima N Y O, **2000**, Evaluation of the impact of household treatment procedures on the quality of groundwater supplies in the rural community of the Victoria district, Eastern Cape. *Water S. A.*, **26**, 285–90.

Leroyer A, Nisse C, Hemon D, Gruchociak A, Salomez J L, Haguenoer J M, Apostoli P, Boffetta P, Landrigan P J, **2000**, Environmental lead exposure in a population of children in northern France: factors affecting lead burden, *American Journal of Industrial Medicine*, **38(3)**, 281-89.

Ling B, **2000**, Health impairments arising from drinking water polluted with domestic sewage and excreta in China, *Schriftenr Ver Wasser Boden Lufthyg*, **105**, 43-46.

Lomborg B, **2001**, The skeptical environmentalist, Cambridge University Press, Chap1, p-22.

Maithani P B, Gurjar R, Banerjee R, Balaji B K, Ramachandran S and Singh R, **1998**, Anomalous fluoride in groundwater from western part of Sirohi district, Rajasthan and its crippling effects on human health, *Current Science*, **74**, 773-77.

Mallo R, Legato A, Erpelding A, Lopez R, Andres M, Tabera A, Murno G, **2001**, Research on water pollution in the Tandil mountain basin, Argentina, *Informacion Technologica*, **12(1)**, 45-50.

Manjapa S, Basavarajappa B E, Desai G P, Hotanahalli S S, Aravinda H B, **2003**, Nitrate and fluoride levels in ground waters of Davanagere taluka in Karnataka, *Indian Journal of Environment and Health*, **45(2)**, 155-60.

Matalas C N and Reiher J B, **1967,** Some comments on the use of factor analysis, *Water Resource Research*, **3(1),** 213-223.

Mathauda S S, Mavi H S, Bhangoo B S, Dhaliwal B K, **2000**, Impact of Projected Climate Change on Rice Production in Punjab (India), *Tropical Ecology*, **41(1)**, 95-98.

Mathur H B, Agarwal H C, Johnson S and Saikia N, **2005**, Analysis of Pesticide Residues in Blood Samples from Villages of Punjab, *Down to Earth*, 2 (June).

Matkar LS and Gangotri M S, **2002**, Physicochemical analysis of sugar industrial effluents, *Journal of Industrial Pollution and Contamination*, **18(2)**, 139-44.

Mattikalli N M, Richards K S, **1996**, Estimation of surface water quality changes in response to land use change: application of the export coefficient model using remote sensing and GIS, *Journal of Environment Management*, **48(3)**, 263-82.

Maurer A M, Stuerchler D, **2000**, A waterborne outbreak of small round structured virus, campylobacter and shigella co-infections in La Neuvevile, Switzerland, *Epidemiology and Infection*, **125(2)**, 325-32.

Maya S, **2003**, Pollution assessment of selected temple tanks of Kerala, *Nature Environment Pollution Technology*, **2(3)**, 289-94.

Meenakumari H R and Hosmani S P, **2003**, Bacteriological examination of ground water samples in and around Mysore city, Karnataka, India, *Nature Environment Pollution Technology*, **2(2)**, 213-15.

Meybeck M, **2004**, The global change of continental aquatic systems: dominant impacts of human activities, *Water Science and Technology*, **49**, 73-83.

Minakshi, Mehra D, Sharma P K, Tur N S, Singh H And Kang G S, **2006**, Assessment, Management and Spatial Distribution of Ground Water for Irrigation in Rupnagar District of Punjab (India), *Journal of Environment Science & Engineering*, **48(2)**, 91-96,

Minakshi, Tur N S, Setia R K and Sharma P K, **2004**, An appraisal of quality of underground water in Fatehgarh district of Punjab, *Indian Journal of Ecology*, **31(2)**, 133.

Mishra D, Mudgal M, Khan M A, Padmakaran P, Chakradhar B, **2009**, *Journal of Scientific and Industrial Research*, **68(11)**, 964-66.

Mishra P C, Behera P C, Patel R K, **2005**, Contamination of water due to major industries and open refuse dumping in the steel city of Orissa – a case study, *Journal of Environment Science and Technology*, **47(2)**, 141-54.

Mittal S K and Verma N, **1997**, Critical analysis of ground water quality parameters, *Indian Journal of Environment Protection*, **17(6)**, 426-29.

Mittal S and Sharma S, **2008**, Assessment of drinking groundwater quality at Moga (Punjab): an overall approach, *Journal of Environmental Research and Development*, **3(1)**, 129-136.

Mohapatra S P, Saxena S K, Ali Arif, **1992**, Occurrence of coliform bacteria in channels receiving municipal sewage, *Indian Journal of Environment Protection*, **12 (7)**, 509-11.

Molden D J, Oweis T, Steduto P, Bindraban P, Hanjra M, Kijne J, **2010**, Improving agricultural water productivity: between optimism and caution, *Agricultural Water Management*, **97**, 528–35.

Momba M N B, Kaleni P, **2002**, Regrowth and survival of indicator micro-organisms on the surfaces of the household containers used for the storage of drinking water in rural communities of South Africa, *Water Research*, **36(12)**, 3023-28.

Mor S, Singh S, Yadav P, Rani V, Rani P, Sheoran M, Singh G and Khaiwal R, **2008,** Appraisal of salinity and fluoride in a semi-arid region of India using statistical and multivariate techniques, *Environmental Geochemistry and Health Official Journal of the Society for Environmental Geochemistry and Health*, Springer

Morisawa M and LaFlure E, **1979**, Hydraulic geometry, stream equilibrium and urbanization in adjustments of the fluvial systems—*Proceedings of the 10th annual Geomorphology Symposium Series*, Rhodes D D and Williams G P (eds.), Inghamton, NY.

Muralidharan D, Rangarajan R, Shankar G B K, **2011**, Vicious cycle of fluoride in semi-arid India-a health concern , *Current Science*, **100(5)**, 638-40.

Murgai R, Ali M, Byerlee D, **2001**, Productivity growth and sustainable in post Green Revolution agriculture: the case of Indian and Pakistani Punjab, *World Bank Research Observer*, **16(2)**, 199-218.

Musa H A, Shears P, Kafi S, Elsabag S K, **1999**, Water quality and public health in northern Sudan: a study of rural and peri-urban communities, *Journal of Applied Microbiology*, **87(5)**, 676-82.

Najafpour S, Alkarkhi A F M, Kadir M O A, Najafpour G D, **2008**, Evaluation of spatial and temporal variation in river water quality, *International Journal of Environment Research*, **2(4)**, 349-58.

Namara R E, Hanjra M A, Castillo G E, Ravnborg H M, Smith L, Koppen B V, **2010**, Agricultural water management and poverty linkages, *Agricultural Water Management*, **97**, 520–27.

Nath K J, Bloomfield S F, and Jones M, **2006**, Household water storage, handling and point of-use treatment, *A review commissioned by International Scientific Forum on Home Hygiene*, (IFH).

Nayak S, **2009**, Distribution inequality and groundwater depletion: An analysis across major states in India, *Indian Journal of Agricultural Economics*, **64(1)**, 89-107.

Neal C, Neal M, Wickham H, Harrow M, **2000**, The water quality of a tributary of the Thames, the Pang, Southern England, *Science of the Total Environment*, **251-252(1-3)**, 459-75.

Nebbache S, Feeny V, Poudevigne I, Alard D, **2001**, Turbidity and nitrate transfer in Karstic aquifers in rural areas: the Brionne basin case study, *Journal of Environment Management*, **62(2)**, 389-98.

Nicholson R V, Cherry J A and Reardon E J, **1983**, Migration of contaminants in ground water at a landfill – a case study, *Journal of Hydrology Netherlands*, **63**, 131–76.

Niederlander H A G, Dogterom J, Buijs P H L, Hupkes R and Adriaanse M, **1996**, UN/ECE Task Force on Monitoring and Assessment- Working programme 1994/1995, **Volume 5**, *State of the Art in Monitoring and Assessment of Rivers*, International Centre of Water Studies, Amsterdam, commissioned by (RIZA).

Niemi G J, Devore P, Detenbeck N, Taylor D, Lima A, **1990**, Overview of case studies on recovery of aquatic systems from disturbance, *Environment Management*, **14(5)**, 571-87.

Niquette P, Servais P, Savoir R, **2001**, Bacterial dynamies in the drinking water distribution system of Brussels, *Water Research*, **35(3)**, 675-82.

Nolan B T, Stoner J D, **2000**, Nutrients in groundwater of the Contenminous, United States, 1992-95, *Environmental Science and Technology*, **34(7)**, 1156-65.

Oeztuerk N, Yilmaz Y Z, **2000**, Trace elements and radioactivity levels in drinking water near Tunebilek coal-fired power plant in Kuetahya, Turkey, *Water Research*, **34(2)**, 704-08.

Okafo C N, Umoh V J, Galadima M, **2003**, Occurrence of pathogens on vegetables harvested from soils irrigated with contaminated streams, *Science of the Total Environment*, **311**, 49 –56.

Olaniya M S and Saxena K L, **1977**, Ground water pollution by open refuse dumps at Jaipur, *Indian Journal of Environmental Health*, **19**, 176–88.

Ongley E D, **1998**, Modernization of water quality programmes in developing countries: Issue of relevancy and cost efficiency, *Water Quality International*, **Sept/Oct-1998**, 37-42.

Oomen J M V, de Volf J, Jobin W R, **1990**, Health and Irrigation, Incorporation of diseases control measures in irrigation, a multi-faceted task in design, construction, operation, *International Land and Reclamation Institute*, Wageningen, The Netherlands.

Patel K P, Pandya R R, Maliwal G L, Patel K C, Ramani V P, **2003**, Suitability of industrial effluents for irrigation around Bharuch and Ankleshwar industrial zone in Gujarat, *Pollution Research*, **22(2)**, 241-45.

Patel L B, Verma V K, Toor G S and Sharma P K, **2001**, Ground water quality for irrigation: Its assessment and management in District Mansa, Punjab (India), *Ecology Environment and Conservation*, **7(3)**, 315.

Patel S K, **1992**, Study of groundwater quality for irrigation in and around Ujjain city (MP), *Journal of Nature Conservation*, **4(1)**, 1-9.

Pathade G R, Molleti V E, Deshmukh A M, **2003**, Enteropathogenic bacterial studies on drinking water in Karad city with reference to drug sensitivity of the isolates, *Nature Environment Pollution Technology*, **2(2)**, 157-62.

Pathak S P, Gopal K, **1994**, Antibiotic resistance and metal tolerance among coliform species from drinking water in a hilly area, *Journal of Environmental Biology*, **15(2)**, 139-47.

Pathak S P, Kumar S, Ramteke P W, Murthy R C, Singh K P, Bhattacharjee J W, Ray P K, **1992**, Riverine pollution in some northern and north eastern states in India, *Environment Monitoring and Assessment*, **22 (3)**, 227-36.

Pawar A C, Nair J K, Jadhav N, Vasundhara, Pawar S C, **2006**, Physicochemical study of ground work samples from Nacharam Industrial area, Hyderabad, Andhra Pradesh, *Indian Journal of Aquatic Biology*, **21(1)**, 118-20.

Pedley S and Howard G, **1997**, The public health implications of microbiological contamination of groundwater, *Quarterly Journal of Engineering Geology*, **30**, 179–88.

Pimentel D, Berger B, Filiberto D, Newton M, Wolfe B, Karabinakis E, Clark S, Poon E, Abbett E, Nandagopal S, **2004,** Water resources: agricultural and environmental issues, *Bioscience*, **54(10)**, 909–18.

Plamondon T, Mills S, **2000**, A practical approach to improving the quality of water used for routine dental treatments, *General Dentistry*, **48(6)**, 682-88.

Postel S L, **1999**, Pillars of sand: can the irrigation miracle last?, W.W. Norton, New York.

Power K N, Negi L A, **1999**, Relationship between bacterial re-growth and some physical and chemical parameters within Sydney's drinking water distribution system, *Water Research*, **33(3)**, 741-50.

Prasad R N, Chandraand R, Tiwari K K, **2008**, Status of Groundwater Quality of Lalsot Urban Area in Dausa District, Rajasthan, *Nature Environmental and Pollution Technology*, **7(3)**, 377-84.

Prater B E, **1975**, The Metal Content and Characteristic of Steel Work Effluents Discharging to the Tees Estuary, *Water Pollution Control*, **74**, 63–78.

Punjab Rural Water Supply & Sanitation (PRWSS) Programme, **2006**, Project Implementation Plan and Operational Manual, Government of Punjab, **22**, p2.

Purandara B K, Varadarajan N, Jayashree K, **2003**, Impact of sewage on ground water quality – case study, *Pollution Research*, **22(2)**, 189-97.

Raghunath H M, **1982**, Groundwater- hydrology, groundwater survey and pumping, rural water supply, Wiley Eastern, New Delhi.

Rahi A S, **2011**, Physicochemical & Microbiological Qualities of Drinking Water, *Research, Analysis and Evaluation*, **2(17)**, 76-77.

Rahi A S, **2011**, Punjab: The Physicochemistry of Groundwater Crisis, *Research, Analysis and Evaluation*, **2(17)**, 92-94.

Raje G B, Muley D V, Mankar D D, **2005**, Analysis of heavy metals in ground water from Lote industrial area in Ratnagir, district (Maharashtra), *Journal of Industrial Pollution Control*, **21(2)**, 381-86.

Rao S M, Mamatha P, **2004**, Water quality in sustainable water management, *Current Science*, **87(7)**, 920-47.

Raucher R S, **1996**, Public health and regulatory considerations of the safe drinking water act, *Annual Reviews of Public Health*, **17**, 179-202.

Rawat M, **2003**, Presumptive coliform count test for the assessment of faecal contamination of two water reservoir of Jodhpur region, *Ecology and Environment Conservation*, **9(1)**, 51-53.

Rayms-Keller A, Olson K E, McGaw M, Oray C, Carlson J O and Beaty B J, **1998**, Effect of Heavy Metals on Aedes aegypti (Diptera:Culicidea) Larvae, *Ecotoxicology and Environmental Safety*, **39**, 41-47.

Reddy R V, **2004**, Managing Water Resources in India: A Synoptic Review, *Journal of Social and Economic Development*, **6(2)**, 176-94.

Reid D C, Edwards A C, Coo Per D, Wilso N E, Mcg Aw B A, **2003**, The quality of drinking water from private water supplies in Aberdeenshire, UK, *Water Research*, **37**, 245.

Richter B D, Braun D P, Mendelson M A, Master L L, **1997,** Threats to imperiled freshwater fauna, *Conservation Biology*, **11(5)**, 1081–93.

Rijal G and Fujioka R, **2001**, Synergistic effect of solar radiation and solar heating to disinfect drinking water sources, *Water Science Technology*, **43(12)**, 155-62.

RIVM, **1992**, The Environment in Europe: A Global Perspective, *National Institute of Public Health and Environmental Protection* (RIVM), Netherlands.

Robertson J B, Edberg S C, **1997**, Natural protection of spring and well drinking water against surface microbial contamination-I, Hydrogeological Parameters, *Critical Reviews in Microbiology*, **23(2)**, 143-78.

Robinson H D, and Maris P J, **1985**, The treatment of leachate from domestic waste in landfill sites, *Journal of Water Pollution Control Federation*, **57(1)**, 30–38.

Rodda J C and Shiklomanov I A, **2003,** World water resources at the beginning of the 21st century, Cambridge University Press, Cambridge.

Ryan P B, Huet N, Macintosh D L, **2000**, Longitudinal investigation of exposure to arsenic, cadmium and lead in drinking water, *Environmental Health perspectives*, **108(8)**, 731-35.

Sabal D, Ashutosh and Khan T I, **2008**, Groundwater fluoride content and water quality in Amber tehsil of Jaipur district, *The Ecoscan*, **2(2)**, 265-67.

Sadeghi G H, Mohammadian M, Nourani M, Peyda M and Eslami A, **2007**, Microbiological Quality Assessment of Rural Drinking Water Supplies in Iran, *Journal of Agricultural Society of Science*, **3(1)**, 31-33.

Saeijs H L F and van Berkel M J, **1995**, Global water crisis: the major issue of the 21st century, a growing and explosive problem, *European Water Pollution Control*, **5(4)**, 26-40.

Saini S P, Chhibba I M and Nayyar V K, **2006**, An appraisal of quality and micronutrient contents of underground irrigation water of south-western region of Punjab, *Crop Research*, **32(1)**, 52-54.

Saleh M A, Ewane E, Jones J, Wilson B L, **2001**, Chemical evaluation of commercial bottled drinking water from Egypt, *Journal of Food Composition and Analysis*, **14(2)**, 127-52.

Samson M, Chawala A, Thukral A, **2010**, Assessment of water quality parameters of the Harike wetland in India, a Ramsar site, using IRS LISS IV satellite data, *Environmental Monitoring and Assessment*, **170(1-4)**, 117-128.

Sauvant M P, Pepin D, **2000**, Geographical variation of the mortality from cardiovascular disease and drinking water in a French small area (Puy de Dome), *Environmental Research*, **84(3)**, 219-27.

Sawyer C N, and McCarthy B L, **1967,** Chemistry for sanitary Engineers, 2nd Ed. *McGraw Hill Book Co.*, New York, 518p.

Saxena S, Upreti D K and Sharma N, **2007**, Heavy metal accumulation in lichens growing in north side Lucknow city, *Journal of Environmental Biology*, **28**, 49-51.

Schaffter N, Parriaux A, **2002**, Pathogenic bacterial water contamination in mountainous catchments, *Water Research*, **36(1)**, 131-39.

Schottler S P, Elsenreich S J and Capel P D, **1994**, Atrazine, alachlor and Cyanazine in a large agricultural river system, *Environment Science Technology*, **28**, 1079-89.

Scott C A, Faruqui N I, Raschid-Sally L, **2004**, Wastewater use in irrigated agriculture: Management challenges in developing countries, p1-10, *Wastewater use in irrigated agriculture: Confronting the Livelihood and Environmental Realities*, Cabi Publishing, Wallingford.

Scott T M, Salina P, Rose K M, Tamplin J B, Farra M L, Koo S R, Lukasik A, **2003**, Geographical variation in ribo-type profiles of Escherichia coli isolates from humans, swine, poultry, beef and dairy cattle in Florida, *Applied Environmental Microbiology*, **69(2)**, 1089-92.

SEGMITE, **1999**, Sustainable development of surface and groundwater resources and exhibition of water purification processes equipments and products, *Society of Economic Geologists and Mineral Technologists.*

Semwal N and Akolkar P, **2006**, Water quality assessment of sacred Himalayan Rivers of Uttaranchal, *Current Science*, **91(4)**, 486-96.

Sharma A, Khurana G S and Dhaliwal G S, **2005**, Awareness of the farmers of Punjab state regarding environment implication caused due to excessive use of Pesticides, *Indian Journal of Ecology*, **32(1)**, 76-78.

Sharma A P, **2001,** Physical, chemical and bacteriological quality of water, *International Workshop on surface/river water quality*, Jointly organized by PCRWR, UNESCO/IHP and German HIP/OHP in cooperation with ICIMOD, Islamabad.

Sharma D R and Minhas P S, **2004**, Soil properties and yield of upland crops as influenced by the long term use of waters having variable residual alkalinity, salinity and sodicity, *Journal of Indian Society of Soil Science*, **52(1)**, 100.

Sharma M R and Verma P S, **2003**, Water quality of springs in Hamirpur area of outer Himalayas, *Pollution Research*, **22(3)**, 369-72.

Sharma M R, Bassin J K, Gupta A B, **2003**, A pollutional profile of Hathli stream in lower Himalayas, *Pollution Research*, **22(2)**, 237-40.

Sharma M, Ranga M M, Goswami N K, **2005**, Study of groundwater quality of the marble industrial area of Kishangarh (Ajmer), Rajasthan. *Nature Environment Pollution Technology*, **4(3)**, 419-20.

Sharma R D, Lal P, Chang S T, Harkar D B, **1992**, Studies on some selected chemical properties of groundwater used for irrigation within Chambal command area, Kota Rajasthan, *Acta Ecologica*, **14 (1)**, 44-47.

Shibu S, Ritakumari S D, Nair N B, **1990**, Environmental inventory and the distribution of inorganic nutrients in a tropical estuary of the southwest coast of India, *Journal of Indian Fisheries Association*, **20**, 59-67.

Shivashankara A R, Shankara Y M S, Rao S H, Bhat P G A, **2000**, Clinical and biochemical study of chronic fluoride toxicity in children of Kheru Thanda of Gulbarga district, Karnataka, India, *Fluoride*, **33(2)**, 66-73.

Shivkumar K and Biksham G, **1995,** Statistical of approach for the assessment of water pollution around industrial areas: A case study from Patancheru, Medak district, India, *Environmental Monitoring and Assessment*, **36,** 228-248.

Shukla S C, Tripathi B D, Mishra B P, Chaturvedi S S, **1992**, Physicochemical and bacteriological properties of the water of river Ganga at Ghazipur, *Comp Physio Eco*, **17(3)**, 92-96.

Sichingabula H M, Nkhuwa D C W, **1998**, Anthropogenic influences on groundwater resources in Lusaka, Zambia, in: *Hydrology in a Changing Environment*, Vol **II**, Wheater H (Ed), Kirby C (Ed), John Wiley & Sons Ltd., UK, 47-60.

Sickman J O, Zanoli M J, Mann H L, **2007**, Effects of urbanization on organic carbon loads in the Sacramento River, California, *Water Resources Research*, **43(W11422)**, 1-15.

Sikdar P K and Banerjee S, **2003**, Genesis of arsenic in groundwater of Ganga Delta – an anthropogenic model, *ENVIS Journal of Human Settlements*, **April 2003**, 10-24.

Simango C, Dindiwe J and Rukure G, **1992**, Bacterial contamination of food and household stored drinking water in a farm-worker community in Zimbabwe, *Central African Journal of Medicine*, **38(4)**, 143–49.

Singh A P and Sakal R, **2001**, Sewage sludge treated soils-Distribution and translocation of micronutrient cations in different plant species, *Sustainable use of Chemicals in Agriculture*, **2**, 22-32.

Singh A, Jully S S, Devi P, Bansal B C and Singh S S, **1962**, An epidemiological biochemical and clinical study in the Bathinda district of Punjab, *Indian Journal of Society of Soil Science*, **27**, 48-53.

Singh B and Bishnoi S R, **1993**, Quality of underground irrigation waters in Barnala Tehsil of Sangrur, district (Punjab), *Indian Journal of Ecology*, **20(1)**, 17.

Singh B and Bishnoi S R, **2004**, Underground irrigation water quality in Muktsar District of Punjab, *Journal of Research, Punjab Agricultural University*, **41(4)**, 442.

Singh B and Sekhon G S, **1976**, Nitrate pollution of groundwater from nitrogen fertilizers and animal wastes in the Punjab, *Indian Agriculture and Environment*, **3**.

Singh B, **1975**, Are fertilizers polluting groundwater?, *Everyday Science*, **20(2)**.

Singh B, **2002**, Pesticidal Contamination of the Environment of Punjab, *Indian Journal of Ecology*, **29(2)**, 189-98.

Singh B, Gaur S and Garg V K, **2006**, Fluoride in drinking water and human urine in Southern Haryana, India, *Journal of Hazardous Materials*, **10**, 1016.

Singh H, Singh R and Singh B, **1990**, Studies in Air, Water and Soil Borne Pollution in Amritsar, *Project Sponsored by Punjab State Council for Science and Technology*, Chandigarh.

Singh I P, Singh B and Bal H S, **1987**, Indiscriminate fertilizers use vis-avis groundwater pollution in central Punjab, *Indian Journal of Agricultural Economics*, **42(3)**.

Singh J, Singh L, Singh S, **1995**, High U-contents observed in some drinking waters of Punjab, India, *Journal of Environmental Radioactivity*, **26(3)**, 217-22.

Singh K P, Malik A, Mohan D, Sinha S, **2004**, Multivariate statistical techniques for the evaluation of spatial and temporal variations in water quality of Gomti River (India): A case study, *Water Research*, **38(18)**, 3980-92.

Sinha R, **2005**, Why do Gangetic rivers aggrade or degrade?, *Current Science*, **89**, 836–40.

Sisti M, Albano A, Brandi G, **1998**, Bacterial effects of chlorine in motile aeromonas species in drinking water supplies and influence of temperature on disinfection efficacy, *Letters in Applied Microbiology*, **26(5)**, 347-51.

Smith C A, Phiefer C B, Macnaughton S J, Peacock A, Burkhalter R S, Kirkegaard R, White D C, **2000**, Quantitative lipid biomarker detection of unculturable microbes and chronic exposure in water distribution system biofilms, *Water Research*, **34(10)**, 2683-88.

Somasekhara R K, Rameshaiah V, Suvarna A C, **2000**, Groundwater chemistry of Channapatana taluk (Bangalore rural district)- regression and cluster analysis, *Journal of Environment and Pollution*, **7(2)**, 101-09.

Somasekhara R K, Venkateswara R L, Padmavathy D, Rambabu C, **1992**, Groundwater quality in Challapalli Mandalam, *Indian Journal of Environment Protection*, **12(5)**, 341-47.

Sood A, Singh K D, Pandey P, Sharma S, **2008**, Assessment of bacterial indicators and physicochemical parameters to investigate pollution status of Gangetic river system of Uttarakhand (India), *Ecology Indicators*, **8**, 709-17.

Sood A, Verma U, Thomas K, Sharma P K and Brar J S, **1998**, Assessment and management of underground waters quality in Talwandi Sabo tehsil of Bathinda district (Punjab), *Journal of Indian Society of Soil Science*, **99**, 421-26.

Squillace P J, Scott J C, Moran M J, Nolan B T, Kolpin D W, **2002**, VOCs, pesticides, nitrate, and their mixtures in groundwater used for drinking water in United States, *Environmental Science and Technology*, **36(9)**, 1923-30.

Srikanth R, **2009,** Challenges of sustainable water quality management in rural India, *Current Science*, **97(3)**, 317-325.

Srikanth R, Vishwanatham K S, Kahsai F, Fisahatsion A, Asmellash M, **2002**, Fluoride in groundwater in selected villages in Eritrea (north east Africa), *Environment Monitoring and Assessment*, **75(2)**, 169-77.

Srinivas T, Kasim M S, Srinivasa R M, **2002**, Study of water quality at solid waste dumping yards in Visakhapatnam, *Journal of Industrial Pollution Contamination*, **18(2)**, 253-65.

Stoddard J L, Jeffries D S, Lukewille A, Clair T A, Dillon P J, Driscoll C T, Forsius M, Johannessen M, Kahl J S, Kellogg J H, Kemp A, Mannio J, Monteith D T, Murdoch P S, Patrick S, Rebsdorf A, Skjelkvale B L, Stainton M P, Traae T, Van Dam H, Webster K E, Wieting J, Wilander A, **1999**, Regional trends in aquatic recovery from acidification in North America and Europe, *Nature*, **401**, 575-78.

Sujatha D, **2003**, Fluoride levels in the groundwater of south-eastern part of Ranga Reddy district, Andhra Pradesh, India, *Environmental Geology*, **44(5)**, 587-91.

Sujatha D, Rajeswara R B, **2003**, Quality characterization of groundwater in the south-eastern part of the Ranga Reddy district, Andhra Pradesh, India, *Environmental Geology*, **44(5)**, 579-86.

Sulochana N, StephenInbaraj B, Selvaraniand K, Thirumurugen V, **1999**, Monitoring, Correlation and Possibilities of Contamination of Groundwater in Thuvakudi Village, Tiruchirapalli District (South India), *Indian Journal Environment Protection*, **19(4)**, 290-95.

Susheela, A.K., **1999**, Fluorosis management programme in India, *Current Science*, **77**, 1250-56.

Szewzyk U, Szewzyk R, Manz W, Schleifer K H, **2000**, Microbiological safety of drinking water, *Annual Review of Microbiology*, **54**, 81-127.

Tanwir F, Saboor A and Shan M H, **2003**, Water Contamination, health hazards and public awareness: a case of the urban Punjab, Pakistan, *International Journal of Agricultural Biology*, **5**, 460–62.

Tarifeno-Silva E, Kawasaki L, Yn D P, Gorden M S and Chapman D J, **1982**, Aqua-cultural Approaches to Recycling Dissolved Nutrients in Secondarily Treated Domestic Waste Waters: Uptake of Dissolved Heavy Metals by Artificial Food Chains, *Water Research*, **16**, 59-65.

Tehounwou P B, Lantum D M, Monkiedje A, Takougang I, Barbazen P, **1997**, The urgent need for environmental sanitation and a safe drinking water supply in Mbandjock, Cameroon, *Archives of Environmental Contamination and Toxicology*, **33(1)**, 17-22.

Thakur J S, Rao B T, Rajwanshi A, Parwana H K, Kumar R, **2008**, Epidemiological Study of High Cancer among Rural Agricultural Community of Punjab in Northern India, *International Journal of Environment Research and Public Health*, **5(5)** 399-407.

Thakur J S, Prinja S, Singh D, Rajwanshi A, Prasad R, Parwana H K, Kumar R, **2010**, Adverse reproductive and child health outcomes among people living near highly toxic waste water drains in Punjab, India, *Journal of Epidemiol Community Health*, **64**, 148-54.

Thevos A K, Kaona F A D, Siajunza M T, Quick R E, **2000**, Adoption of safe water behaviors in Zambia: comparing educational and motivational approaches, *Education for Health*, **13(3)**, 366-76.

Thurman R, Faulkner B, Veal D, Cramer G, Meiklejohn M, **1998**, Water quality in rural Australia, *Journal of Applied Microbiology*, **84**, 627-32.

Tibbetts J, **2000**, Water World 2000, *Environmental Health Perspectives,* **108(2)**.

Tiller K G, **1992**, Urban soil contamination in Australia, *Australian Journal of Soil Research*, **30**, 937-57.

Tiwana N S, Jerath N, Singh G and Singh R, **2007**, Pesticide Pollution in Punjab: A Review, Asian Journal of Water, *Environment and Pollution*, **6(1)**, 89-96.

Toze S, **1999**, PCR and the detection of microbial pathogens in water and wastewater, *Water Research*, **33(17)**, 3545-56.

Tripathy J K, **2003**, Groundwater hydrochemistry in and around Bhanja Bihar, Ganjam district, Orissa, *Pollution Research*, **22(2)**, 185-88.

Umar R and Absar A, **2003**, Chemical characteristics of groundwater in parts of the Gambhir River basin, Bharatpur District, Rajasthan, India, *Environment Geology*, **44(5)**, 533-45.

UNEP, **1996**, Characterization and assessment of groundwater quality concerns in Asia-Pacific region, *Environment Assessment Report,* British Geological Survey.

United Nations Population Fund (UNFPA), **1997**, Population and Sustainable Development: Five Years after Rio, UNFPA, New York, 1-36.

Vahter M, Berglund M, Akesson A, Liden C, **2002**, Metals and women's health, *Environmental Research*, **88(3)**, 145-55.

Van der Hoek W, ul Hassan M, Ensink J H J, feenstra S, Raschid-Sally L, Munir S, Aslam R, Ali N, Hussain R, Matsuno Y, **2002**, Urban Wastewater: a valuable resource for agriculture-a case study from Haroonabad, Pakistan, *International Water Management Institute*, Colombo.

Van der Straten J W H, Semmekrot S, Maasdam R and de Haan H, **1998**, Regional Water System Report in the picture, *H₂O*, **15**, 17-18.

Van Leeuwen F X R, **2000**, Safe drinking water: the toxicologists approach, *Food and Chemical Toxicology*, **38**, S51-58.

Van Rooy P T J C, **1995**, Towards comprehensive water management in The Netherlands (2) bottlenecks, *European Water Pollution Control*, **5(6)**, 33-40.

Vander slice J and Briscoe J, **1993**, All coliforms are not created equal: a comparison of the effects of water source and in-house water contamination on infantile diarrheal disease, *Water Resources Research*, **29**, 1983-95.

Varadarajan N and Purandara B K, **2003**, Hydrochemical characteristics of groundwater: a case study, *Ecology and Environment Conservation*, **9(3)**, 253-62.

Verma V K, Sharma P K, Seta R K, **2007**, Ionic composition and hazards of poor quality waters for irrigation in southwestern part of Punjab, *Hydrology Journal*, **30(3&4)**.

Vijay Shankar P S, Kulkarni H, Krishanan S, **2011**, India's groundwater challenge and the way forward, *Economic and Political Weekly*, **XLVI(2)**, 37-45.

Vlek P LG, Denich M, Martius C and Rodgers C, **2006**, Water, sanitation, hygiene and diarrheal diseases in the Aral Sea area (Khorezm, Uzbekistan), *Journal of Ecology and Development*, **2(43)**, 81-90.

Volk C J, Hofmann R, Chauret C, Gagnon G A, Ranger G and Andrews R C, **2002**, Implementation of chlorine dioxide disinfection: Effects of the treatment change on drinking water quality in a full-scale distribution system, *Journal of Environment Engineering Science*, **1**, 323–30.

Ward R C, Loftis J C and McBride J B, **1986**, The "data-rich but information-poor" syndrome in water quality monitoring. *Environmental Management*, **10(3)**, 291-97.

Welch P, David J, Clarke W, Trinidade A, Penner D, Berstein S, McDougall L and Adesiyun A A, **2000**, Microbial quality of water in rural communities of Trinidad, *Review of Panam Salud Publications*, **8(3)**, 172–80.

WHO, **1991**, Guidelines for drinking water quality, Vol **2**, *World Health Organization*, CBS Publishers and Distributors, New Delhi.

WHO, **1996**, Guidelines for drinking water quality, Vol **2**, 2nd Ed., *World Health Organization*, Geneva.

Witmer M C H, **1995**, Information needs for policy evaluation, 1994, *Proceedings of the international workshop Monitoring Tailor-Made-I*, Beekbergen, the Netherlands, p55-61.

Woolf A, Wright R, Amarasiriwardena C, Bellinger D, **2002**, A child with chronic manganese exposure from drinking water, *Environmental Health Perspectives*, **110(6)**, 613-16.

Wyatt C J, Fimbres C, Romo L, Mendez R O, Grijalva M, **1998**, Incidence of heavy metal contamination in water supplies in Northern Mexico, *Environment Research*, **76(2)**, 114-19.

Yadav A K, Jain P K, Sharma J, **2003**, Assessment of ground water quality of Behror tehsil of Alwar District (Rajasthan), *Aquaculture*, **4(2)**, 265-70.

Yang C Y, Cheng M F, Tsai S S, Hsieh Y L, **1998**, Calcium, magnesium and nitrate in drinking water and gastric cancer mortality, *Japanese Journal of Cancer Research*, **89(2)**, 124-30.

Yang C Y, Chiu H F, Cheng M F, Tsai S S, Hung C F, Lin M C, **1999b**, Esophageal cancer mortality and total hardness levels in Taiwan's drinking water, *Environmental Research*, **81(4)**, 302-08.

Yang C Y, Tsai S S, Lai T C, Chun-Fang Hung C F, Chiu H F, **1999a**, Rectal cancer mortality and total hardness levels in Taiwan's drinking water, *Environmental Research*, **80(4)**, 311-16.

Young P, **1996**, Safe drinking water- a call for global action, *ASM News*, **62**, 349-52.

Zacheus O M, Lehtola M J, Korhonen L K, Martikainen P J, **2001**, Soft deposits, the key site for microbial growth in drinking water distribution networks, *Water Research*, **35(7)**, 1757-65.

Zakova Z, Berankova D, Koekova F, Kriz P, Mlejnkova H, **1993**, Investigation of the development of biological and chemical conditions in the Vir reservoir 30 years after impoundment (Review), *Water Science and Technology*, **28(6)**, 65-74.

Zamaxaka M, Pironcheva G and Muyima N Y O, **2004**, Microbiological and Physicochemical Assessment of the Quality of Domestic Water Sources in Selected Rural Communities of the Eastern Cape Province of South Africa, *Water S. A.*, **30**.

Zamxaka M, Pironcheva G and Muyima N Y O, **2004**, Microbiological and physicochemical assessment of the quality of domestic water sources in selected rural communities of the Eastern Cape Province, *South Africa Journal of Bioc. and Microbiology*, **30(3)**, 333-40.

Zomer R, Trabucco A, van Straaten O, Bossio D, **2006,** Carbon, land and water: a global analysis of the hydrologic dimensions of climate change mitigation through afforestation/reforestation, Research Report 101, International Water Management Institute, Colombo.

Zwieg R D, Morton J D, Stewart M M, **1999**, Source Water Quality for Aquaculture: A Guide for Assessment, The World Bank, Washington.